FIRE

FLOOD

PLAGUE

Australian writers respond to 2020

Edited by Sophie Cunningham

VINTAGE BOOKS

Australia

VINTAGE

UK | USA | Canada | Ireland | Australia
India | New Zealand | South Africa | China

Vintage is part of the Penguin Random House group of companies whose addresses can be
found at global.penguinrandomhouse.com

First published by Vintage in 2020

Versions of some of these essays were originally published on the website of
the *Guardian* Australia from July to September 2020.

Excerpts from the poem 'Grief' by Martin Johnston from the collection *Beautiful Objects*
(Ligature, 2020) reprinted with the permission of the Estate of Martin Johnston.

Cover image by sbayram/Getty Images
Cover design by James Rendall © Penguin Random House Australia Pty Ltd
Typeset in 12.5/18 pt Bembo by Midland Typesetters, Australia

Printed and bound in Australia by Griffin Press, part of Ovato, an accredited
ISO AS/NZS 14001 Environmental Management Systems printer

 A catalogue record for this
book is available from the
NATIONAL
LIBRARY National Library of Australia
OF AUSTRALIA

ISBN 978 1 76104 040 5

This project is supported by the Copyright Agency's Cultural Fund

CULTURAL FUND

penguin.com.au

MIX
Paper from
responsible sources
FSC FSC® C009448

Contents

CONTENTS

Timeline*

2019

September

The bushfire season began in August and by 10 September, eighty bushfires are burning in Queensland and sixty in New South Wales. The Gondwanan rainforests of Australia are on fire for the first time in recorded history.

December

Sydney is covered in smoke from the beginning of December until January 2020.

26 Fires burn in the Stirling Ranges, Western Australia.

30 There are three active fires in East Gippsland, Victoria.

31 The Currowan, Charleys Forest and Clyde Mountain fires join in New South Wales, forming a massive fire front that burns through more than 257,000 hectares. Cobargo, New South Wales, burns. Fire reaches Mallacoota, Victoria, causing more than 4000 people

to shelter on the beach. Health authorities in Wuhan, Hubei province, China, reveal that they have been treating several cases of viral pneumonia.

2020

January

1 Tens of thousands march in protest in Hong Kong, almost one year since an extradition bill was proposed. The Canberra air quality reading is the worst on the planet: twenty-six times levels considered hazardous to human health.

2 Prime Minister Scott Morrison visits Cobargo, but locals make it clear he isn't welcome.

3 Fire on Kangaroo Island, South Australia, burns 170,000 hectares: one third of the island. The Australian Navy evacuates residents and tourists from Mallacoota.

4 Penrith, New South Wales, records temperatures of 48.9°C, making it the hottest place on Earth at the time.

5–6 Canberra records the worst air quality reading on the planet again, causing public buildings, government offices and businesses to close.

6 The federal government commits $2 billion to a national bushfire recovery fund.

7 It's estimated that one billion animals have been killed in the bushfires. That figure will be revised up to three billion by the middle of the year.

9   The Australian Bureau of Meteorology declares 2019 Australia's hottest and driest year on record.

11 China announces the first coronavirus-related death.

14 Fires rage across New South Wales and temperatures reach 49°C. Premier Gladys Berejiklian declares a state of emergency. Air quality in Melbourne is the worst in the world and remains at dangerously high levels for several days.

16 An extensive (and successful) operation is undertaken to save the Wollemi Pines from the Gospers Mountain fire. It has been burning since October and is the biggest forest fire in Australian history.

20 A hailstorm causes extensive damage across Canberra. The storm cell also hits parts of New South Wales and Victoria, and creates $638 million worth of damage in its wake.

21 The Morrison government activates its coronavirus emergency response plan.

23 The Chinese government imposes a lockdown of Wuhan and cities in Hubei province.

25 Australia records its first case of the novel coronavirus.

30 The World Health Organization declares the coronavirus outbreak a global emergency.

February

7–10 Two thirds of the fires in New South Wales have been extinguished, but there are now heavy rains and flash floods along the east coast.

8   Severe Tropical Cyclone Damien hits Dampier–Karratha region in Western Australia.

11 WHO announces 'COVID-19' as the official name of the disease caused by the novel coronavirus.

27 The Australian government declares that the coronavirus outbreak will become a global pandemic.

28 The Australasian Fire and Emergency Service Authorities Council (AFAC) reports that a total of 17 million hectares and 3094 houses have been burnt, and thirty-three people have died over the course of the 2019–20 bushfire season. By August, satellite estimates will suggest that 30 million hectares had burnt.

29 There are 86,152 cases and 2924 deaths from coronavirus globally. There are now twenty-four cases in Australia.

March

The third mass coral bleaching on the Great Barrier Reef in five years occurs. The 2020 bleaching is the most widespread ever recorded.

2 Australia records its first COVID-19 death.

11 WHO declares the COVID-19 outbreak a global pandemic.

15 All international arrivals to Australia are required to self-isolate at home for fourteen days.

16 The ASX 200 posts its biggest daily percentage fall on record: 9.5 per cent.

18 The Australian Governor-General declares a 'human biosecurity emergency' in the country: a first. Travel advice is upgraded to Level Four: Do Not Travel. Tasmania closes its borders.

19 2700 passengers disembark from the cruise ship
    Ruby Princess in Sydney, despite several passengers
    falling ill with flu-like symptoms while onboard.

20 Australia closes its borders to all non-citizens and
    non-residents.

23 The National Cabinet's lockdown measures come into
    effect: hospitality venues move to takeaway only; gyms,
    cinemas and other entertainment venues close.

24 South Australia closes its borders.

25 Australia bans its citizens from travelling overseas.

26 The New South Wales–Queensland border closes.

27 Non-essential travel into remote Indigenous communities
    in the Northern Territory is banned. Scott Morrison
    announces that all returning international travellers will
    be required to self-isolate in hotels for fourteen days.

29 The National Cabinet limits public gatherings to
    two people. The curve begins to flatten.

31 There are now 768,967 COVID-19 cases and 40,867
    deaths globally, with 4359 cases and eighteen deaths
    in Australia.

April

5   Hard border restrictions are introduced in Western
    Australia.

8   The Chinese government lifts lockdown in Hubei
    province, and its capital, Wuhan.

12 The Murray and Darling Rivers join for the first time
    in two years.

30 3,105,637 COVID-19 cases and 224,677 deaths globally.
    There have been 6746 cases and 90 deaths in Australia.

May

23 46,000-year-old Juukan Gorge ancient rock shelters are blown up by mining company Rio Tinto in the Pilbara.

25 An African-American man, George Floyd, is killed by police in Minneapolis, Minnesota. While hundreds of civilians are killed by police each year in the United States, the rate of fatal police shootings among African Americans is close to three times higher than that of any other ethnicity. In some parts of the US, police kill Black people at a rate six times higher than they kill white people.

26 Black Lives Matter protests begin in Minneapolis.

31 There have now been 5,941,223 COVID-19 cases and 366,363 deaths globally, with 7185 cases and 103 deaths in Australia.

June

6 Black Lives Matter protests take place in Sydney, Melbourne and Brisbane, calling for an end to Aboriginal deaths in custody. By this date, 432 Aboriginal people had died in custody since the Royal Commission ended in 1991.

20 A monitoring station in the Siberian town of Verkhoyansk registers a record high of 38°C north of the Arctic Circle. June is declared the Earth's warmest June since records began in 1850.

30 The Chinese government imposes new criminal secession and sedition laws in Hong Kong, an hour before the twenty-third anniversary of the handover of the former British colony to Chinese rule in 1997.

30 There have now been 10,185,374 COVID-19 cases and 503,862 deaths globally. 7834 cases and 104 deaths in Australia.

July

4 Nine public housing towers with 3000 residents are ordered to go into an immediate five-day police-enforced hard lockdown in Melbourne. Residents are banned from leaving their apartments for any reason.

7 Stage Three lockdown is re-imposed in Melbourne.

8 The Victoria–New South Wales border closes.

10 Queensland opens its borders.

17 Homes on New South Wales's Central Coast fall into the sea after a series of tidal surges and erosion events.

19 Restrictions ease for eight out of nine public housing towers in Melbourne. Masks become mandatory in metropolitan Melbourne and Mitchell Shire.

31 There have now been 7,112,724 COVID-19 cases and 663,597 deaths globally, and 16,905 cases and 196 deaths in Australia.

August

4 Two massive explosions rock Beirut. The event is linked to ammonium nitrate that had been stored, by the Lebanese government, without proper safety measures for six years. More than 200 people die, 6000 are injured and 300,000 are left homeless. COVID-19 cases surge in the wake of the bombings.

5 Melbourne's lockdown is upgraded to Stage Four. Regional Victoria is put into Stage Three lockdown.

7 Queensland closes its borders again.

31 There have now been 25,156,046 COVID-19 cases and 844 639 deaths globally, with 25,746 cases and 652 deaths in Australia.

September

2 The Australian economy is officially declared to be in recession.

6 The Stage Four lockdown for metropolitan Melbourne is extended until at least 27 September.

7 The remaining two Australian media correspondents working out of China are pulled out of the country, signalling a dramatic escalation of diplomatic tensions.

10 102 fires burn 4.4 million acres across the West Coast of the United States.

13 Restrictions ease across regional Victoria but remain in place in metropolitan Melbourne. Premier Daniel Andrews unveils his 'roadmap' to reopening Victoria.

27 Daniel Andrews announces the easing of some restrictions as part of Victoria's roadmap.

30 There have been 33,249,563 confirmed cases of COVID-19, and 1,000,040 deaths globally, with 27,044 cases and 875 deaths in Australia.

★ Note: These statistics are intended as a guide only. Figures vary between organisations, and are revised over time by these organisations.

Introduction

SOPHIE CUNNINGHAM

April in Melbourne is always glorious but through most of the autumn of 2020, between the hours of five and six, there was an exquisite clarity to the rose-gold sheen of the sky. Less pollution was one theory, as the world slowly shut down. Whatever the reason, I basked in the glow through my windscreen and sat talking to an author about their contribution to this anthology. The car speakers were a nice way to take a call. More relaxing than Zoom. Safe. This was before I accepted the need to invest in Zoom Pro, a microphone and earbuds. This was before I realised this was only the first lockdown of what would likely be several. I would be sitting in enclosed spaces for the next six months and that was unlikely to be enough to contain the virus. Not by a long shot.

I'd wanted to do other things with my car this year. Specifically, I'd wanted to drive it to Bermagui on New South Wales's South Coast, to spend time with family over the new year.

But as the new year dawned – violent, smoky – there were bushfires to contend with, then air quality so dangerous my Canberran loved ones were trapped in their house. Soon enough there were hailstorms smashing into their workplaces. More fires, floods, then the plague. On it went. We understood that summer fires followed by late summer floods were considered to be part of the cascading effect of climate change. We understood that deforestation led to an increased likelihood of pandemics, but frankly, people can't look every which way all at once and anyway it seemed that the genie was out of the bottle, the cat was out of the bag, the tipping point had tipped and now we were in the territory of the unprecedented, the territory of pivoting, the territory of grief and loss.

However, as a writer who has spent much of the last couple of years making gloomy pronouncements about the state of the world, there is one particular thing I didn't see coming: the collapse of the federation and the dissolution of the State into its component states and territories. The impossibility of driving to Bermagui. I tried to get there once, twice, thrice, but first the Alpine Way was on fire, then the Princes Highway. For two months after that, various sections of road were closed between Victoria and New South Wales. At the end of March the roads were opened but Stage Three lockdown restrictions came into play. I planned to go in mid-July but by then, even if I'd been allowed to travel, the border had been closed.

I'm worried it might be years before I can spend time with my family again. Years before I get to see the forest of spotted gums lining the road that sweeps into Bermagui; or the sleepy town's pelicans; or the pale-blue armies of soldier crabs pouring out onto the mudflats when the tide is low; or have the chance to walk the piece of old highway that has been turned into a nature reserve and left to fall into the sea. The Road, I liked to call it, in homage to Cormac McCarthy.

Many years are described as the year everything changed and perhaps that framing of 2020 is not ultimately useful. But it's fair to say that the year 2020 will be remembered, at least by those who lived through it, as the year the human race fell off a cliff. It's unclear how long that descent will take, how deadly it will be, and what shape we'll be in when we land.

I am spectacularly privileged to have been spared — because of class, time of life, accidents of fate — from everything other than the existential crisis posed by the COVID-19 crisis at this point. My main experience is one of extreme claustrophobia. And if I had to pick the moment when I felt the most personal distress in a year which offered a cornucopia of anxieties I'd say that it was during the bushfires. I continue to reserve my greatest fears for what will happen when global temperatures rise 2, 3, 4 degrees. Which is all by way of saying I didn't pay enough attention when I was sitting in a campground over New Year's and a fellow camper read out the headline. 'CHINESE AUTHORITIES TREAT DOZENS OF CASES OF PNEUMONIA OF UNKNOWN CAUSE,' then commented, 'Sounds like a worry.' I was trying to find enough reception to track what was happening in Mallacoota, where 4000 people stood on the beach under blood-red skies as fires roared through East Gippsland.

The thing about change is that it makes no difference what we think we know, or don't know; how many events we can take in at once, or respond to; how we might imagine we can keep safe. **Jane Rawson** speaks eloquently to this notion of safety. Is it achievable? Is it the point? And while we can entertain fantasies of self-sufficiency and national agency and whether we're talking about carbon credits or COVID-19, we all breathe the same air, as **Jennifer Mills** points out, and we're all 'bit players in a complex ecology'. Borders and bluster meant nothing to the weather, or to the virus.

So, four months after I'd been camping, there I was, sitting in my car, working on this book, talking about what global economic collapse might look like. Wondering what would happen with China. Wondering how many lives would be lost in the United States before Trump (I refuse to use his honorifics) was managed (or thrown) out of office. Wondering how realistic a vaccine seemed. Not very, I thought, but at the same time I found it impossible to imagine quite what that looked like: a world where there was no vaccine. And to be honest, while I've often spoken and written about what environmental collapse might feel like, what I was quickly learning was that while much that has happened this year was predicted, the experience was continually surprising. **Tom Griffiths** describes the time between the fires and the pandemic as one during which, 'people spoke courageously of "the new normal", but did not yet understand that "normal" was gone . . . their masks were still in their pockets' and soon enough they'd need them again. There is no returning to normal, and, as **Alison Croggon** writes, that can only be a good thing. Normal was not going well.

It is impossible to comprehensively document a year as profound as 2020 has been, but the writers herein have all charted different aspects of the terrain. Some essays are deeply personal, others have chosen to step back and offer us some perspective. Some of the difficulties and traumas that have erupted are too all-encompassing to be simultaneously lived through and documented, so there are some notable absences (homeschooling, working in a hospital, having a parent with COVID-19 in aged care, being shut into public housing with very little warning and no access to food or medical supplies).

I, like many of the writers in this collection, have found myself writing essays once, twice, thrice as we've progressed from bushfire and smoke-choked skies to the early days of the pandemic to where I sit now: working from my bedroom, in Stage Four lockdown, past the anxiety of the first stage and into the exhaustion of what is becoming a marathon. **James Bradley** writes bravely of the experience of his mother dying in the early days of the pandemic. **Delia Falconer** and her family fall through time as the year unfolds. **Brenda Walker** listens to a sonic version of the virus, **Rebecca Giggs** asks if our senses are changing, heightened, by the experience of plague. **Kate Cole-Adams** embraces new ways of relating as we're kept at a distance from each other, reaching across borders and through time zones using technology. **Kirsten Tranter** mourns as she watches the ashes of our great forests rain down upon us.

The timeline we've included reminds us all how much has happened in the first nine months of a year in which a global pandemic has met environmental unravelling – confronted here by **Joëlle Gergis** – and fuelled an increasingly unhinged political scene. In this year of political chaos **Lenore Taylor** gives us some good news: fact-based reporting may be making a comeback, and if we follow the facts wherever they lead, the media will be strengthened at a time it is most needed.

Revisions have been required. Just before publication, Gergis reworked her devastating essay to reflect that the estimated number of Australian animals lost in the bushfire was not one billion but an unimaginable three. **Omar Sakr** finished a draft of his essay before the explosions tore through Beirut in August, then had to consider if words could do any justice to that horrendous event. **George Megalogenis** had to make a call on the potential impact of a cut in immigration, both as a result of COVID-19 and rising xenophobia in an increasingly unstable environment. He, like **Richard McGregor**, looked to the past to make his case, McGregor using history to help us understand what we might expect of our relationship with China in the months and years to come. **Tim Flannery** asks if there isn't something to be learnt from our government's response to the coronavirus. What can be achieved if we move from a place of denial to political will and action? **Gabrielle Chan** shows us the ways that rural communities are coming together to form new supply chains in a contribution that provides some welcome news, giving us a glimpse of the possibilities present in times of upheaval. Her exploration of economic models based on community suggests a return to a more collegial, hierarchical infrastructure, a possibility embraced in many of the essays. To do this, we need to push through what **Kim Scott** describes as neo-liberalism's dog-eat-dog mentality to an understanding and an embracing of our Aboriginal heritage, an essential part of which is that we both listen to and adopt the Uluru

Statement from the Heart. I write of the ways in which rural Australia is disenfranchised by the failing management of its life-blood: water. This connection between our current economic and social structures and the problems we are facing recurs. **Jess Hill** traces the links between patriarchy, colonialism, the bushfires and the pandemic. **Melissa Lucashenko** reminds us that the idea of a safe Australia is an illusion the country's First Nations have never enjoyed. The physical and economic constraints imposed by the pandemic on the population have been imposed on Indigenous Australians from the year 1788. White settlers brought pandemic with them and the colonial (and patriarchal) project unleashed both cultural and environmental damage we are all having to live with today. When Australians rallied in support of the Black Lives Matter movement, when they tried to make explicit the relationship between the murder of African-Americans at the hands of police and the appalling record of deaths in custody for Indigenous Australians, they were told that street marches were not safe. But being a person of colour is never safe in this nation and the responses to both the rallies and to those the media chooses to humiliate as the virus has spread is colour coded. These links between our past and present are delicately, movingly, described and explored by **Billy Griffiths**. **John Birmingham** vividly steps us through the othering of not just a people but of our entire nation, as fire engulfed our forests and the eastern seaboard. We were rendered to the rest of the world as TV news footage. Two dimensional. A nation in crisis. A nation to be pitied.

Christos Tsiolkas is bracing. He reminds us that those of us who imagined that our lives of travel and privilege, of relative ease, might last were dreaming. But, as I'm sure all who pick up this anthology will attest, few of us have slept well this year. As **Melanie Cheng** enunciates, the language of horror movies, the language of nightmares, is what is haunting us. Dreaming and hope don't come easily. **Nyadol Nyuon** shares her hard-won knowledge: we need to resurrect our inner lives if we are to survive in this new world. Optimism sometimes seems impossible but finding hope and transforming it, constructively, into our lived reality is the only work worth doing right now. The question then becomes, what do we want to happen next? What is it that we are hoping for?

This place of sickness

BILLY GRIFFITHS

The year began in haze. We felt relatively safe where we were on the New South Wales South Coast, despite the blanket of smoke. The Currowan blaze had passed through here a month earlier, leaving little left to burn. I could see for hundreds of metres into the once thick bush. Charred black trunks rose from scorched earth towards auburn treetops. The trees at the fringe of the small township were stained pink with fire retardant: the eerie residue of an airdrop that ultimately saved my partner's family house.

We were not safe, of course, as so many Australians learnt this fire season. There was nothing predictable about the fires that raged from winter into spring into summer, smouldering into autumn. Not only was fire reaching into ecologies unfamiliar with flame, but it was looping back on itself, with fallen leaf litter fuelling new spot fires. This was a new breed of fire: a biome pushed to its limits by anthropogenic climate change. I later read how the smoke from the south-east fires had carried east across the seas to choke cities in New Zealand and, after circumnavigating the globe, bruise the skies of Perth.

In the final days of 2019, the Princes Highway closed in both directions. We lost power and mobile reception. ABC local radio became our only connection to the outside world. At night, we listened as dead trees creaked and crashed nearby. The resident kangaroos huddled in the garden, drinking thirstily from the birdbath, grazing the dry lawn. Two months later one came back to rest and die in the shade of the water tank. My father-in-law buried her in the ash nearby. By winter, the mob had gone. The township feels lonely without them.

We evacuated the region in the early hours of 2 January, seizing on a narrow window when the roads were clear, following a police escort through a corridor of still-burning bush. Fallen powerlines smouldered on the roadside. Paint had melted off the signs. Blackened gum leaves fluttered from the lurid orange sky.

The smoke stayed with us all the way back to Melbourne. The fug of tragedy was just as thick. Some of our friends lost family in these fires; others lost houses; we all lost places we love.

For First Nations peoples, this sense of loss is acute. At Deakin University, where I teach, Aboriginal flags were lowered to half-mast in mourning for the destruction of Country, with all its complex meanings and associations. As Gabrielle Fletcher, Director of the National Indigenous Knowledges, Education, Research and Innovation (NIKERI) Institute wrote, 'To lose Country, in this way, is a distinct, messy kind of grief.'

I can remember my first day of 2020 without smoke. It was late in January, and I was on Flinders Island, researching an essay about the Bassian Plain, the vast land bridge that connected Tasmania and the mainland during the last ice age. I drove to a high point near Emita and gazed west into the teeth of the wind, savouring the fresh air. For the first time I could make out the distant islands that pock the Strait: the mountains and ridges that once must have guided travellers across the plains.

Below me lay Wybalenna, the dismal concentration camp where more than 200 Tasmanian Aboriginal peoples were herded and detained from 1833 to 1847. The thin, rotting wattle-and-daub walls that housed them are long gone. All that remains is the historic chapel and a sombre graveyard. It was known to those who lived there as 'a place of sickness'.

On lungtalanana/Clark Island, to the south of Flinders Island, a team of researchers recently uncovered a 41,000-year record of Aboriginal fire management. It is a powerful reminder that fire has always been a part of life in Australia. When Cook voyaged here 250 years ago, he was struck by the burning coastline. He wrote in his journal that he had come upon a 'continent of smoke'. *This continent of smoke.* In the midst of our savage summer – the black summer, the forever fires – his words seemed prescient. And yet the fires he saw were nothing like those we experienced. The smoke told a story of life, ceremony and interconnection. It rose from campfires, fire signals and cool, cleansing, managed burns. It was healthy and local, not continental. Cook was being closely watched on his voyage along the coast, and the Sea Country he charted was already intimately known and embedded with culture and Law.

In the wake of the latest bushfires, there has been a surge of interest in Aboriginal peoples' fire knowledge. The gentle smoke of firestick farming or 'cultural burning' is once again being seen as a positive force, perhaps even a saving grace. Yet, as Indigenous fire practitioner Victor Steffensen argues in his book *Fire Country*, Aboriginal environmental knowledge is holistic: it cannot simply be an addendum to existing management regimes. Cultural anthropologist Tim Neale observes that all too often, non-Indigenous individuals and institutions have 'sought to understand Aboriginal fire knowledge and practices without empowering or even engaging with Aboriginal peoples'.[1] Steffensen calls for a more equal relationship. His book is as much about power (or powerlessness) as it is about fire: 'I'd love for the government to effectively jump in the passenger seat and let us drive for a change.'[2]

As the novel coronavirus consumed the bushfire crisis we became witnesses to an example of what happens when Aboriginal leaders take the keys from government. As early as January, many Aboriginal community-controlled health services began planning for COVID-19, seeking advice from First Nations communities in Canada, sharing information and developing a strategy to protect their people against an as-yet-undeclared pandemic. In February, a national taskforce, the National Aboriginal Community Controlled Health Organisation (NACCHO), was formed, meeting twice weekly. The Northern Land Council released public health messages in eighteen different languages. About the time that panic buying was making national headlines, Aboriginal communities quietly began closing borders: first in the Anangu Pitjantjatjara Yankunytjatjara Lands and then in Cape York and parts of Western Australia. By the end of

June, fewer than sixty cases of COVID-19 had been recorded among Aboriginal and Torres Strait Islander peoples, representing 0.8 per cent of all Australian cases.

The Aboriginal health sector's response to COVID-19 has been coordinated, evidence-based and often far ahead of federal and state governments. It is a small insight into what can happen when Indigenous communities are empowered; when their leaders have control over their peoples' fates. As Darumbal/South Sea Islander journalist Amy McQuire reports, part of the reason for this swift and decisive action is that Aboriginal and Torres Strait Islander communities have borne the brunt of past viruses, from the H1N1 swine flu pandemic in 2009 to the threat of Zika in the tropical north. They also carry with them historical memory of the devastation that disease can bring. Not only from the 'Spanish' (actually American) influenza outbreak of 1919, in which some Aboriginal communities had a 50 per cent mortality rate, but also from earlier outbreaks, such as smallpox, which was an ally in the conquest of Australia.

I wonder how the COVID-19 pandemic will shift Australians' historical imaginations. Will it allow us to better grasp the role of disease in invasion? Will it help us appreciate, with the full depth of compassion, the enormity of the smallpox epidemics in the eighteenth and nineteenth centuries?

Disease killed some 80 to 90 per cent of the population in parts of eastern Australia. Let's be clear: this was and remains the single greatest demographic catastrophe in Australian history.

According to some estimates, the Aboriginal population fell from some 750,000 or perhaps even 1.5 million in 1788 to only about 60,000 in the 1920s. Although violence was ever present, this was mostly due to introduced diseases, such as smallpox, measles and influenza, as well as venereal and respiratory diseases. The numbers are vague because of how swiftly disease spread.

The 1789 smallpox epidemic is particularly vivid in the historical sources. The new pathogen, known to Eora and Darug as gal-gal-la, engulfed the societies of Sydney fifteen months after the British landed. It ripped through densely populated areas and provoked terror and disbelief. There was no cure. It extinguished entire language groups, killed warriors in the midst of resistance and disproportionately affected Elders, stealing cultural knowledge as well as lives. It left those who remained wondering where everyone had gone, coming to terms with haunting absence and mournful silence. And it travelled far ahead of the frontier, so that when European explorers eventually moved inland, they were greeted by the pockmarked faces of survivors.

The source of the 1789 smallpox epidemic remains disputed. There were no recorded cases among the First Fleet, so it could not have been incubated in the bodies of the invaders. And yet the time between arrival and outbreak is surely too close to be a coincidence. Over the past century many scholars have determinedly sought an alternative source for the spread of smallpox. A popular theory, first put forward by microbiologist John Burton Cleland in 1912, links the epidemic to recorded outbreaks in Sulawesi in the early 1780s, with the disease transmitted by Macassan seafarers in the north and then sweeping across the continent, arriving in Sydney in 1789. This hypothesis conveniently washes British hands of culpability, and it remains vigorously contested. We have learnt a lot about the spread of disease these past months and there was plenty of social distancing happening between Sydney and Sulawesi.

A more likely scenario, proposed by historian Craig Mear, is that the disease was transmitted by smallpox scabs – or formites – clinging to old clothing and blankets held in a ship's store. Blankets were the colonial currency of welfare, and the gift of a blanket could have been a deadly transaction. The question is: was infection inadvertent? Or was it a half-conscious, even sometimes deliberate act of biological warfare? The debate continues among historians, although there is no doubt that British commanders used smallpox as a weapon against First Nations peoples in North America.

As Australia entered lockdown in March, many historians began publicly drawing connections with earlier outbreaks of disease and the looming 250th anniversary of Cook's voyage along (not around) the Australian coastline. A month later, on 29 April, the Deputy Chief Health Officer of Victoria, Annaliese van Diemen, tweeted a similar comparison:

> Sudden arrival of an invader from another land, decimating populations, creating terror. Forces the population to make enormous sacrifices & completely change how they live in order to survive. COVID19 or Cook 1770?

Her three-line tweet sparked a flurry of conservative outrage and bad puns. COOK LINE A STINKER read the front page of the *Herald Sun*. Home Affairs Minister Peter Dutton made a rare public appearance in the wake of the *Ruby Princess* debacle to call for van Diemen to be sacked. Prime Minister Scott Morrison dismissed the tweet, noting that van Diemen 'wouldn't get the job as chief historian'. (If only there were such a role!)

But the comparison is apt. Regardless of his role in conquest, Cook is emblematic of British arrival – an idea that has been reinforced by government spending on anniversaries and statues. Two hundred and fifty years after his landing on Australian shores, are we ready to confront this complex historical legacy? Anniversaries should inspire reflection and debate. They should incubate new histories.

Part of the 'Great Australian Silence' has been a willingness to see disease as inevitable and apolitical. During this pandemic we have witnessed every country in the world confront the same biological phenomenon with dramatically different consequences. The global story we are watching unfold is not only about microbes; it is also about culture, politics and history. The spread of disease is not without responsibility.

In Australia, disease is often invoked to qualify the culpability of the invaders. But, as historian James Boyce reminds us, the British well understood the link between disease and conquest, yet took few measures to protect vulnerable populations. In Melbourne, for example, the highest rates of mortality occurred *after* the initial act of dispossession, when Aboriginal peoples had lost their traditional food supplies and were concentrated together in places with poor sanitary conditions. At Wybalenna, where Tasmanian Aboriginal peoples were confined for fifteen years, the death rate was more than 70 per cent. As Boyce writes, 'The point is not the one made by former Prime Minister Paul Keating in his Redfern Park speech of 1992, that "we brought the disease", but rather that Europeans created the conditions in which diseases flourished and did almost nothing to ameliorate these.'[3]

The destruction wrought by introduced diseases cannot be explained by biology alone. It was, and continues to be, a matter of public health.

On that smokeless day in January, I walked down the ridge towards the historic site of Wybalenna and spent some time in the restored chapel, listening to the wind wail in the rafters. The visitors' book is full of comments from the descendants of survivors, who go there regularly to remember, reflect and to feel connection with their ancestors. The recurring theme in the comments is the sentiment that 'this must never happen again'. And yet, I cannot help but compare it with the offshore detention centres of today.

On the ridge above the site is a grainy black-and-white photograph printed on a metal sign. It was taken in 1858 at Oyster Cove, where those who survived this 'place of sickness' were removed after Wybalenna was abandoned in 1847. It shows nine Tasmanian Aboriginal people sitting outside a wooden house while a dog plays in the foreground. Their gaze is defiant. The only interpretation offered on the sign is a solemn declaration: 'The morality of a nation remains impaired until that nation deals honestly with its past.'

Looking into their faces, still haunted by the savaged coastal forests, I am reminded of Gamilaraay and Yawalaraay journalist Lorena Allam's response to the bushfires: 'We know what it feels like to lose everything.'

Billy Griffiths is a writer and historian at Deakin University. His latest book, *Deep Time Dreaming*, was named Book of the Year at the 2019 NSW Premier's Literary Awards.

The year of lethal wonders

JOHN BIRMINGHAM

The chicken shack was nearly an hour's walk through Seoul in the sub-zero night, but they served up damn good chicken and dangerously cheap beer, and we agreed the risk of becoming lost and freezing to death on the street was worth it. My son, Thomas, spent his early years in Canberra, and he does not feel the cold like I do, routinely sleeping with his bedroom windows wide open through the winter. But on this night even he swaddled up with multiple layers of hoodies, scarves and so much Korean puffer-wear that we were less men than giant, shambling marshmallows in search of the dirty bird.

There was, as well, a quiet pleasure to be had from the killing cold. When we had flown out of Australia a few days earlier, the whole of the sky was smeared a smoky orange ochre, and the familiar steam-press humidity of summer in the subtropics had evaporated under a furnace blast of dry heat from the heart of the continent. It felt good to shiver and contemplate the lot of everyone we'd left behind, especially as we drummed greasy fingertips on painfully distended tummies full of spicy chicken meat. But enjoyment would pass.

We were nearly two weeks in Seoul, and by the time we made our last pilgrimage to the chicken joint, unimaginably vast expanses of Australian bushland and even rainforest had not just burned but been atomised inside a conflagration burning at planetary scale. My long habit when travelling overseas of never reading Australian news came to nothing, because the firestorm at home was headline news all over the world. A giant video wall across the street from our hotel, two or three storeys high, pulsed with apocalyptic visions of rich, white refugees hip deep into the ocean, having fled the black tsunami of ash, smoke and flame. International news sites refreshed constant updates from the fire front, and from the eerily deserted streets of Sydney and Melbourne. Scientists estimated the number of native animals killed to be in the millions, then hundreds of millions. Then more than a billion.

This was horror painted on a vast, burning canvas. But worse were the smaller, more intimate stories of fireys and farmers who, having heard the dying screams of burning koalas, could never unhear them. And through it all, the prime minister was inexplicably missing in action, until it turned out he was actually just missing in Hawaii.

Watching from Seoul, seven and a half thousand kilometres away, it seemed to me that it could not go on. Something would have to give. But it would, of course, go on for months.

To be so far away during those days of the summer holocaust was distancing in more than one way. At night, sitting in the bar in my hotel, staring out over the Blade Runner landscape of Seoul's Hongdae district, I could doom-scroll Twitter for apocalyptic visions of hellfire consuming Australia. Sometimes I could glance down nineteen storeys to see the same images on those giant screens just above street level, where they competed for eyeballs against the pixel storms of StarCraft and Counter-Strike competitions from the gaming café across the street, or the even more surreal imagery of the giant animated cartoon mascots that are ubiquitous in Korea. Australia's horror was abstracted away, made alien and rendered into meaningless colour and movement.

If this seems harsh, I can only offer my congratulations. You have reacquired your sense of privilege.

How many famines in Africa, ground wars in Asia or catastrophic oil spills in the Gulf of Mexico have you contemplated from afar without suffering visceral connection? You may have felt something akin to sympathy, but more likely you were processing the idea of sympathy because the words or images which carried the knowledge of those terrible events to you seemed to come freighted with a moral imperative that you care about them.

At least until you scrolled away, went out for coffee, or realised that Netflix had dropped the next season of your favourite show and it was time to get your binge on.

I remember feeling very strongly as I watched Australia burn from Seoul and Hong Kong and, months later, from Rome and Milan and Paris, that we were the Other now. Our turn had come. And while the world might gin up some long-range empathy for the plight of a dying koala or baby kangaroo, we the peeps were probably shit out of luck. Just like those poor bastards we exiled to our Pacific gulags, who drew the shortest straw when their fates became dependent on our generosity of spirit. Or the Eora people, whose long watch over the southern shores of Botany Bay was closed out by Arthur Phillip's order to a young Watkin Tench to lead a punitive raid against the tribe to spread 'an exemplary terror among the natives'.

Sometimes your luck just runs out, and a hard truth of human nature is that we really only care about what's close. White Australia was a lucky country for so long because the gift of distance was history's lack of interest in our affairs. But that absence of concern for a tiny outpost, removed from the centre of things, can just as easily turn to indifference and genuine disdain when fate turns against us. From afar it was possible to see with the cold objectivity of foreigners just how unflattering a picture we made for any who cared to look. An immensely privileged dominion occupied by a small number of deeply selfish people, suddenly confronted with the consequences of inaction.

For a few months there you could finally see the world accelerating towards the existential discontinuity of irreversible, devastating climate change. No more projections. No theories. No modelling or arcane math. The future had arrived. It was not evenly distributed. It had exploded into the real on the eastern edge of the Australian continent.

Meanwhile, sitting on a beach in Hawaii was our doughy, aggressively know-nothing prime minister, infamous for carrying a big lump of coal onto the floor of parliament and fondling the same with the puckish joy of a man-sized Billy Bunter in possession of a large, unexpected jam doughnut.

To the beach he went, while his land and his people hurtled towards the burning pit.

Perhaps, sipping umbrella drinks and mugging for happy snaps with similarly footloose bogans, Scott Morrison was himself subject to the distancing effect I felt all around me in South Korea and Hong Kong, that deeply human flaw that gaming journalist and Twitter savant David Milner describes as the inability to conceive as real any reality different to our own lived experience. Hong Kong in particular afforded a novel perspective on the subject-object divide, as smoke from nearby pro-democracy riots drifted into the bar where we sat watching smoke from a series of megafires blanket the streets of Sydney.

The drinkers of Hong Kong's mid-level district seemed no more inconvenienced by the actual particulates contaminating their imported beers than they were by hi-def imagery of incinerated biomass contaminating the city where some of that beer had originated. The riots that weekend were a few miles away. The megafires another world.

But of course otherness is an illusion; comforting, anaesthetising, but illusion nonetheless.

By that time, eighty per cent of Hong Kong's population had been teargassed at some point in the previous six months, and every day drew them closer to the end of the city's freedom, many years ahead of the prescribed return to complete control by the mainland.

It is perhaps a passing irony of 2020, the year of metastasising ironies, that Bejing's grip on its recalcitrant entrepot should have been strengthened by the emergence of the novel coronavirus from the wet markets of Wuhan. But nobody said the end times would be boring, and if there is subtext to this year of lethal wonders, it cautions against narrative complacency. I had assumed a thousand years ago in February that a world with its own sorrows to attend to would eventually tire of ours, even while the bush burned and the reef turned white and died. But a funny thing happened on the way to the apocalypse, and that congenital flaw of the human heart and mind – our inability to connect intimate subject to distant object – would be undone.

By distancing.

The virus may yet kill us all, one way or another, but having killed at least half a million human beings at the time of writing, having emptied the streets of Rome and the tall towers of Manhattan, it scattered us like chaff before the tempest – or like aerosolised droplets of infected sputum, should you prefer a more clinically befitting metaphor. To our varied holes we bolted, those of us who could, leaving behind those who couldn't because they did not have the means to flee and we still needed them to drive trucks, stack shelves, flip burgers and remind us of a simple truth that heroism is often average people doing normal things when everything else is falling apart. But in that falling apart, for a little while at least, it felt like we were all together.

Racism, selfishness and moments of mad panic flared up here and there, and often all at once in the toilet paper aisle at the local supermarket, but as the human race contemplated our shared mortality it did seem, for one brief shining moment, that our inability to imagine much beyond our lived experience was suddenly irrelevant because everyone's lived experience was now the same. The virus was coming for us.

Of course, we have moved on from that. Politics in particular abhors a vacuum and has rushed back into the space cleared by COVID-19. That's not completely unreasonable. In the US, which lurches closer to catastrophic failure every hour the tangerine Chimpenführer sits on the throne, only politics can save them from something much worse. Here, the working truce between state and federal governments and the major parties must inevitably fracture when we turn to the forever question of who is going to pay for everything. (Spoiler, it won't be the rich.) And in the Hobbesian realm of geostrategy, the uncertainties and instabilities occasioned by the pandemic can only increase those deep, tectonic pressures running along the fault lines between civilisations. This would have happened whether China bungled its initial response to the virus or not.

But home again, exiled to my pleasant writing room with the nice view over the remnant forest that runs down to the river, I can only wonder, what next.

The weather in Siberia is ominous.

Months of extreme heat have baked the Russian Arctic and thawed gigatonnages of permafrost. The frozen tundra warms even as the coronavirus spreads. They are, for now, unrelated but coincident planetary comorbidities. Warm weather is not unheard of in the Arctic. Summer days are long up there, allowing for a lot of solar energy to hit the ice caps. As the underlying seasonal temperatures creep up, however, those occasional heat waves that do occur will add their thermal power to a much higher baseline.

And now twelve million acres of Russia's coldest wastelands are on fire.

They burn while we distract ourselves with contention over other things. They burn while you read this. They will burn tomorrow when you have found some new distraction. Eventually, we can but hope, winter will return and they will stop burning, but by then our turn will have come around again, and the Arctic Circle will be just that little bit less arctic.

What ancient viruses slumber in those cold sediments waiting for a flicker of warmth to loose them into the world again, or lurk in the Amazon, or the jungles of West Africa, needful only of the woodsman's axe to carve a path to freedom?

Reducing everything to dualism is another of our faults, but this does seem to be a binary moment. Either we are all in this together, or we are doomed to return to a state of nature, the never-ending war of all against all. It was Prospero at the end of *The Tempest* who warned us against illusions and otherness, lest we see dissolved our 'cloud-capped towers, the gorgeous palaces, the solemn temples and the great globe itself'. I fear, I really do fear, that either we heed the old sorcerer, or, as he further warned, we shall 'leave not a rack behind'.

John Birmingham wrote for magazines before publishing *He Died With A Falafel In His Hand*, working for *Rolling Stone*, *Wisden* and *The Independent* among others. He won the National Prize for Non-Fiction for *Leviathan: An unauthorised biography of Sydney*.

Black flowers: Mourning in ashes

KIRSTEN TRANTER

December 2019. The Sydney Harbour Bridge disappears into a pall of smoke. The usual measurements for air quality seem almost absurdly useless for knowing how to act or how to make sense of this crisis: ten times, eleven times hazardous levels. Beaches are darkened by ash in the water. Our house in the inner-west smells of burning, even after being closed all day. The smoke alarm sounds periodically, piercing the night, waking the children. Fragments of burnt forest fall in the city, bringing to mind the black flowers imagined by Emma Craven, the murdered daughter in the 1985 nuclear noir BBC series *Edge of Darkness*.

I watched *Edge of Darkness* as an anxious teenager on the verge of political awakening, worried about nuclear war. My scepticism was already formed, but it was given new contours, new imaginative extent, by this story – a political narrative underwritten by mythic force.

Re-watching the show recently, much of it felt painfully relevant. It is a story about how humans are wrecking the world, and this alone explains why it returns to me so insistently. *Edge of Darkness* is a doomsday tale, but it is also, crucially, a narrative about grief. The death of the planet told as a story about personal mourning: this, I suspect, is what brings to mind images of the stricken face of Detective Ronald Craven, played by a grim Bob Peck; the dark, toxic caves; the crackle of the Geiger counter; the laugh of the cynical, mad American agent. At the beginning of the series, Craven's daughter Emma is shot to death on their front doorstep. He uncovers a giant conspiracy as he doggedly investigates her murder. Emma turns out to have been involved with an activist group trying to bring attention to the dangers of nuclear power. Her death was a political assassination.

Emma appears as a chatty ghost from time to time. In one conversation with her father she explains that the earth will protect itself against the threat that humans pose, and that the consequences for humans will be bad. She calls this philosophy 'Gaia', the name of the ancient Greek goddess of the earth. The screenwriter, Troy Kennedy Martin, seems to have been influenced by the thinking of climate scientist James Lovelock, who proposed a similar theory of earth as a single, interconnected organism.

Once upon a time, Emma says, millions of years ago, life was threatened by a long ice age, and the planet brought forth black flowers to cover the surface of the world. These dark petals absorbed the sun's warmth, allowing life to re-emerge. 'That is the power of Gaia,' she claims. Those black flowers will bloom again, Emma predicts, soaking up the increasing heat from the atmosphere, hastening the warming of the planet, melting the ice caps and devastating the human population. 'If man is the enemy, it will destroy him,' she tells her father.

At the end of the series, Craven stands alone on a cold hill. He does not turn into a tree, which Martin envisaged in his original version of the screenplay, but he is tree-like. Black flowers bloom against the snow, perfect round petals shivering in the wind.

I think of these flowers when I read about the phenomenon of dark snow, where particles of pollution fall on the Arctic tundra, reducing the landscape's albedo, which is the ability of the snow to reflect light and heat back into the atmosphere. I think of them when I read about forms of toxic algae that thrive in warmer waters. And again when I learn about Gamba grass, an invasive species that burns hotter and faster than any native flora, and also grows far more quickly, upending the traditional cycles of fire and native regrowth in the landscape of the Northern Territory. And again when the massive blazes to the south generate thunderstorms over the dry earth, storms of lightning that bring no rain, just more fire.

One December afternoon in Sydney's inner-west, the sun appears as a neon disc, electric orange-pink against the greyish-brown smoke that covers the sky. The haze flattens the sun's rays so that we walk through an eerie false twilight, empty of shadows.

'You can see the sun!' my four-year-old son exclaims.

I quickly cover his eyes and explain that even though it doesn't hurt to look at the sun like this, it still hurts your eyes on the inside, where you can't feel it.

Later, I stand in the front yard in the sinister evening light, trying not to look at the sun as it sets. A white flake drifts past, uncannily like snow. But it is not snow. It is the opposite of snow, snow's horrifying evil twin. It is a piece of ash, I realise, in a long, awful moment of recognition. More of them spiral downwards, so light that gravity struggles to bring them to earth. They have been blown here from the fires many miles away.

Another association bothers the edges of memory. I remember standing on Second Avenue in New York's East Village on September 11, 2001, after watching the Twin Towers burn from my corner of First Street and First Avenue. The parade of shell-shocked people – men in their suits, women in their sensible office heels – making their way uptown from Ground Zero, all dusted with ghostly white. Their shambling gaits, their blank survivors' faces.

As the days passed in New York, we learnt more about the composition of this white dust: toxic chemicals and concrete and paper and cremated people. Our apartment was right on the border of the area where you could request a free air purifier from the city administration. We never applied for ours.

The fragments of disaster that fell on New York ranged from microscopic particles to dust to snowflake-sized scraps to bits of paper, charred around the edges, the writing still legible. Everything could be made into pieces, this event announced, dissolved into its constitutive elements one way or another. One of my friends emerged from the World Trade Center subway station that morning of 9/11 to find a section of aircraft fuselage partially blocking the exit. There was a piece of an aeroplane there, he said to me. But no one was looking at the piece of aeroplane, they were all looking up, he said, and when he looked up too he saw the metal on the sides of the towers melting, turning into liquid.

Where is it from, I wonder, the ash falling on Marrickville? Which of the many fires that dot the map, shifting closer together every time I refresh the screen?

The gravity-defying ash seems to have travelled not simply through space but also through time, challenging not only geographic scale but temporal scale, too. *This ash is the remains of ancient rainforests that have never been burned*: some version of this remark has become commonplace, repeated by friends. It is tinged with fantasy, I know, this idea of the untouched forest, the place of innocent nature defiled by human action. And yet it it is hard to escape the feeling that the ash has come to us from some primordial era, from a forest and a time before human industry, before the Anthropocene itself. *We existed before you,* the ash seems to say. *With these fires we will displace you.*

In the anguished performance of public mourning for the forests, for the koalas and marsupials and baby bats, the unimaginable number of three billion animals, I begin to sense the displacement of a vast inarticulate sorrow at what colonial settlement has wrought for these past 200 years and more. Grief for all those trees and forests cleared, that habitat destroyed, those animals hunted to extinction. The people murdered. The genocide.

Climate scientists voice their concern that some forests have been devastated beyond recovery, leaving a destructive carbon deficit. How long can these fires go on, we all wonder? *Soon there will be nothing left to burn*, friends joke, half-joke. But there is so much left to burn.

The morning after the first December ashfall, the courtyard is covered in fine fragments: white, grey and charcoal black. My son left a plastic toy knife on the ground and when I pick it up there is a dirty penumbra around its shape, a pixelated shadow of ash, like the chalk outline around a murdered body.

Traces, cinders. I think of Jacques Derrida, who valued these figures above all others in his thinking about the problem of presence and absence, but Derrida is no help.

A leaf rests in my hand, a pale grey ghost transformed by extreme heat, impossibly fragile. It is like an inverse fossil: organic matter turned to ash, one touch away from dust. Miraculously, it retained this form as it travelled. It seems to gather to itself metaphors, similes, measures of comparison that might help to contain the unthinkable enormity, the giant present and future catastrophe it represents.

On the concrete there is another leaf, which has tightly rolled as it burned. A piece of bark coming apart into fibres, like some delicate woven textile on the verge of coming undone. A shiny blackened crisp of something, dotted with tiny bubbles. It breaks when I lift it, and the disintegration of these things, the perfect leaf, the unidentifiable charred piece of forest, makes the raw tide of grief rise.

I feel a compulsion to gather these fragments, to keep them intact, a blind ritual of mourning. A collection of relics builds on our wooden sideboard.

I want to document these objects, to photograph them, to

craft an image from light and shadow that will preserve them. My husband pulls out his big camera with the good lens, able to capture the startling, distinct textures of the pieces. The piece of bright, glassy rounded bark, so shiny it almost looks wet with lacquer. Beside it, the dry matte-grey leaf. The depthless black of the fibrous bark, marred by a smear of white ash.

What will we do with this reliquary of charred leaves and bark, I wonder, when we move in four weeks' time, preparing to return to the United States? It is difficult to imagine transporting these things safely, these objects the wind has carried to us from such a distance. They encode survival and extinction all at once, death masks of the forest. They look so beautiful on the camera's bright little screen, and I am ashamed of my impulse to make them into pictures. It seems to betray or belittle the tragedy they represent, the sheer brutal choking violence of it. What happens to that violence when their material form is made aesthetic? The anger, the injustice, the fury? What is the role of such mourning within the sphere of action that must be taken in order to make any impact on the climate catastrophe?

I treat the fragments with tenderness, with caution, understanding that I am acting like a person in shock, performing mindless comforting actions. The habitus of grief. I hope there will be no more of them to collect. I understand that this hope is baseless.

We wait for the southerly, for the cool change, but the wind is hot and carries the smell of burning. Smoke speeds across the face of the orange moon.

I think of Emma Craven's sad, enigmatic smile as she watches her father destroy himself in the course of his search for answers, his passionate sense of purpose in a world that is already dead for him without her. What makes Emma's fatalism so morbidly appealing? I suspect it is almost the only form of hope left to me, a depressive soul at the best of times, disillusioned politically, agreeing in a dejected way with Frederic Jameson or whoever said that it is easier for most people to imagine the end of the world than the end of capitalism. And yet I know that belief in the power of imagination is the closest thing I have to faith or hope.

Recently a friend posted a link to an article about a list of novels in a genre affiliated with climate fiction described as 'hopepunk', a term that made me snort with scorn. What could be less punk? But then I listened again to the Sex Pistols' punk anthem 'God Save the Queen' and heard anew its stark brutal sense of possibility amid destruction. Now I think maybe Emma's black flowers are its perfect floral emblem. Toxic, powerful, darkly incandescent as the rage that animates this grief.

The song is famous for its statement that there is no future, but there's more to it. The future is over for the fascist regime, but not everyone else. 'We're the flowers in the dustbin,' Johnny Rotten sings, 'the poison' that will annihilate the machine. 'We're the future.'

Poisonous garbage flowers of the future. That might be an image of hope I can get behind.

Let the reliquary crumble, I tell myself. Lay down the blackened wreath. 'Grief breaks the heart and yet the grief comes next,' wrote the poet Martin Johnston. It is relentless. Perhaps that drive is where the answer will be found to the question of what to do next, what force will shape the step away from the grave.

'Some lemon morning in a wash of rain,' Johnston's poem continues, 'a brand-new horror comes to call again and write a footnote to expunge the text'.

Months later, at the start of fire season in California, the early morning reeks of smoke. Fine white flecks of ash dust the railing and planks of the deck, the plastic swimming pool, the falling-apart wooden chairs in the yard. It's just past dawn, so early I could almost try to convince myself that this is all a dream, a memory of Sydney. But the ash is different: fire-bleached and powder-fine, offering no after-image of whatever it was that burned. Fires surround us, yet more historically unprecedented fires. A wall of pale grey obstructs our view of San Francisco, across the water; the next day it is impossible to see further than a few blocks in any direction through the haze and dirty yellowish light. We were here for the fires that shut down the Bay in 2018, and it is like that, but now we are in COVID-land, on day one-hundred-and-forty-something of the official shelter-in-place order, and the libraries and indoor play gyms I visited back then with the bouncing small child are all closed. We unbox the air filter. The feeling of concrete settles in my lungs. I can't tell anymore if it is panic or ash sediment.

I think of the electric tension between Emma's message of doom and her father's unstoppable determination, the craggy tree that faces down the black flowers. I want to know: what will take root on this scorched earth?

Kirsten Tranter publishes fiction and criticism. She is the author of three novels, including *Hold*.

Living in the time of coronavirus

DELIA FALCONER

The sense of time-slip begins during the summer megafires. Walking my children home from school in Sydney under a red sun I have the nagging feeling, beneath my anxiety, that I've seen this close orange light before. Then I remember. My father made our family nativity set out of pumpkin-coloured cardboard, topped with a skylight of red acrylic. The sideboard lamp cast the same uncanny glow onto baby Jesus and his shadowless entourage.

Three months later, in early March, my partner and I are driving the twins down the South Coast through green dairy country to isolate from the coronavirus. 'Does the sky seem particularly blue to you?' he asks, as we look up the valley. 'I'm having a "severe clear" moment.' A dark joke between us: pilots used the term to describe a sky of perfect visibility on the morning of September 11. With most planes cancelled, there are no bright contrails in the usually busy flight path above the escarpment. The air is alert and tender. It occurs to me that we haven't seen a sky like this since our own childhoods, near the beginning of the Great Acceleration, when the indicators of human activity on the 'planetary dashboard' began their upward surge.

For this whole first week we feel strangely elated. When we last stood in the back paddock of this Airbnb at Christmas, we watched a huge pyrocumulus build above the Currowan megafire, forty kilometres to the south, as white ash fell onto the brown grass around us. When it sucked the oxygen from the air, we fled. But over the next months, as the eastern seaboard of the country burned, the fires spared this pocket of rainforest tucked beneath the Barren Grounds Reserve. Huge rains fell in February, the owner tells me, as she keeps her social distance, overflowing the creek and tumbling huge boulders down its bed. The flowering gums in the bush around us give off a scent of honey, and the calls of tiny birds electrify the air.

Stop the World: I Want to Get Off. My mother used to jokingly quote the title of this musical to me when I was little. 'Stop the world,' its hero said whenever things went wrong. I've been feeling this way for the last few years. I can't help feeling relieved that the world has at least paused now.

Like Melville's Ishmael, I am a natural 'Isolato'. An only child and a writer, I've been training for this moment all my life. But it's not just me taking pleasure in a world temporarily stilled. A friend texts me video his brother has filmed in Rome of a man walking naked on an empty bridge across the Tiber. On Twitter, people tweet and retweet images of trafficless Venetian canals, water so calm you can see fish in their powdery aqua depths. Footage recorded by an unseen man walking the city's empty *calles* and bridges goes viral: 'Bizarre!' he concludes at its end. I email my publisher. When was the last time the canals were this magically glassy – when the Brownings were restoring the Ca'Rezzonico, or in Casanova's time? He replies that he is amazed by the grand scale of these uncrowded public spaces. 'You look at old pictures of Venice and St Mark's Square is almost empty – really, they're different places entirely. The stones of Venice come into their own.'

We give the twins a phone and a snake bandage and they explore the paddocks while we work. Their saxophone and flute lessons continue by Skype. At the lip of the cliff at the end of the garden they watch a lyrebird go through its repertoire of other birds' songs with the fixed concentration of a flasher, tail feathers lifted stiffly over its back. I chalk this up as a win for homeschooling.

Online, we watch events unfolding at a distance. Our neighbourhood Facebook group fills with queries about whether it is possible to train safely in shared apartment gyms, if balayage will be considered an essential service, and whether favourite restaurants will deliver. Newspapers report toilet paper hoarding while conspiracies circulate that buses are descending on small towns to clear the shelves. A German friend says there is a word for this in her language: *Hamsterkaüfe*, rodent purchases.

Back home, in an inner-city apartment with no verandah or yard, it is harder. The children need constant help with their Year 3 school assignments, which are distributed across Seesaw and two different versions of Google Classroom. My partner and I work fitfully at our own computers. When a hairy, shirtless man in the block next door starts playing guitar to the neighbourhood kids from the outdoor walkway, I feel like putting my head out of my study and yelling, 'This isn't Italy, mate!' In the bathroom, my daughter switches from singing 'Happy Birthday' to the Sex Pistols' 'God Save the Queen' while she washes her hands. As the death toll in Italy and Spain rises, I read an interview with a virologist who says he is so attuned to particles of virus that his view of every surface is 'pixelated'.

The threat level rises after we learn that a cruise ship, the *Ruby Princess*, has been allowed to disgorge its passengers, unchecked, to wander through the city. Many further their travel, some untraced. I search for clips of the plague ship arriving into Delft in Herzog's *Nosferatu the Vampyre* and instantly regret it. On the neighbourhood Facebook page, pleas burn and flare out as bargaining turns to acceptance. Where is it possible to still find good charcuterie? Is there a mobile nail service that might visit? Does anyone know anything about agencies for fostering dogs? When crowds mass at nearby Bondi in the autumn heat, ignoring restrictions on large groups, all the beaches in the Eastern Suburbs are ordered to close. The council fences off the exercise equipment in our park, which has also been busy, and heaving joggers take over the paths.

I begin to walk with the children to the city instead, to the empty Botanic Gardens, along the least used of the long historic staircases that plunge down the Woolloomooloo Bay cliff. 'Look,' I say to them, as we stand across from where the foundations for the huge extension of the Art Gallery of NSW are being dug. 'You will remember this.' I was their age when I used to pass the half-built 'Lego apartments' which now occupy the clifftop behind us, as my parents drove to my grandmother's each weekend. My father always repeated the rumour that the missing heiress Juanita Nielsen, who fought the developers trying to destroy the historic Potts Point streetscape on which the apartments were built, was buried in the foundations.

'Look,' I say again, as we drive back one afternoon from helping my mother, who still lives in the family home on the North Shore. 'You will never see the Harbour Bridge as empty again in your lifetimes.' 'Is it really empty? Was it this empty when you were growing up?' my son asks, as we pass only a handful of cars on its ten-lane approach. Yes, I tell him, there were two million fewer people in the city, which was still in parts a Victorian ruin. We fly over the Cahill Expressway, above the Quay, where no ferries or charter cruise ships move, and on, through an empty Woolloomooloo, as their childhood melts into mine.

Ecologists write about shifting baseline syndrome, in which each new generation takes the reduced world of creatures it grows up with as the starting point for measuring the loss of natural abundance. But now, as if in a time machine, we seem to be glimpsing a new past, or at least an alternative baseline – a world in which we're less intrusive. The viral images of swans and porpoises in the Venetian canals have been debunked but they are still circulating on my social media feed because the desire to believe in them, or their promise of a different future, is so strong. As if anticipating new freedoms, a long green tree snake is photographed climbing the curved Deco brickwork of an apartment in nearby Potts Point. A friend in lockdown in a

Spanish village reports, 'There are numerous owls everywhere around us ... and they have really become gaily, proudly, shamelessly symphonic during humanity's lockdown!'

Throughout April, infections in Australia continue to rise, a significant percentage of them from the *Ruby Princess*. At least children don't seem to be dying, we say quietly to the godparents on the phone. In the Botanic Gardens, a security guard now moves on people sitting on the grass or squatting by the pond, where eels patrol the perimeter as if in parodic imitation. Police cars drive along the footpaths of our local park to fine people sitting in the sun, but do not stop the joggers. A historian friend gives the children an hour-long lesson on FaceTime and tells them that the first year of Sydney's settlement began with plague, asking them to imagine the Aboriginal people who died from smallpox soon after the First Fleet's arrival.

On our neighbourhood Facebook page, people post pictures of sunsets, rooftop concerts, and video graphics from a Belgian–Dutch study showing that joggers' and cyclists' breaths can infect from a distance of five metres, which they then accuse the group administrators of deleting. Although I vowed I would never have one in the house, I have a small coffee pod machine delivered and look out onto a street filled with couriers' vans. A group of us who used to meet a decade ago revive our fortnightly games night through Zoom and we find ourselves speculating, over wine, that we are in for a much longer lockdown than we first expected. I text a friend whose mother has just died. 'I'm sitting out the back behind our house looking at a planeless sky,' he texts back. 'It feels like the end of the world.'

Meanwhile, on my feed, images keep appearing of turtles laying eggs on empty Indian beaches and flocks of wild goats besieging a Welsh town. But nature, it seems, is not getting as much respite as I hoped. Airlines are lobbying to loosen their carbon lowering obligations once borders reopen, while Trump's southern wall, dividing the fragile and biodiverse wildlife of the borderlands, reaches completion almost unnoticed. In Australia, logging continues in the stands of eucalypts that survived the fires and the government announces a review of environmental 'green tape'. I read a study predicting wildlife collapse from climate change will happen suddenly and quickly – a cliff edge rather than a slippery slope – and the future rushes in again.

Over the last few years I have been writing about our entry into the Anthropocene and the way, even as the present accelerates, we are haunted by the deep past, in the form of the fossil carbon we have released into the atmosphere. I have found it unbearable to think too far into my children's future or answer their questions about what they might be when they grow up because I do not believe it will resemble the present in any way. After the fires, I was bracing for another disaster, although I can't help feeling that this one is only the grace note, with its intimations of alternative futures that later catastrophes will not give us.

What I didn't expect was that time would become so kaleidoscopic. At the beginning of lockdown, friends on Facebook participated in a viral 'challenge', posting photos of their younger selves. It is a bittersweet thing to look at them now. All those soft young faces.

Delia Falconer is an award-winning novelist, short-story writer, journalist, non-fiction writer and critic. In 2018, she was the winner of the national Walkley Pascall Prize for Arts Criticism. Her books include *The Service of Clouds*, *The Penguin Book of the Road* and *Sydney*. Delia is a Senior Lecturer in Creative Writing at the University of Technology, Sydney.

Flames

ALISON CROGGON

Where do I begin? Andy Williams' schmaltzy song earworms into my head. It came out in 1970 as the theme to the film *Love Story*, which I never saw. I vividly remember the girls at school talking about it: their obsession with it, and my envy because I hadn't seen it and wouldn't see it and couldn't join in. But maybe I'm misremembering?

I was eight years old the year *Love Story* came out, which seems too young for the memory, which was surely from high school. Wikipedia tells me it aired on US television in 1972. Was it broadcast here? I started high school when I was ten years old, so maybe it is plausible. Maybe the movie was released in Australia later than everywhere else? I don't remember.

Memory is like that. Events overlay and confuse each other, some disappear altogether. As you get older, it becomes harder to remember the order in which things happened. Everything rushes through the narrow aperture of the present into the wide, dimensionless ocean of the past. And that ocean just keeps getting bigger.

New Year's Eve 2019 seems like years ago. I loathe New Year's Eve: it's the one night of the year I'm guaranteed to be home. As the clocks ticked over into 2020, I spent the evening glued to Twitter, watching as bushfires raged across eastern Australia threatening township after township: Mallacoota, Batemans Bay, Merimbula, Cobargo, Conjola Park, Corryong, Cudgewa, Sarsfield.

Like anyone raised in the country, bushfire days – those days when you wake up and smell the north wind roaring down from the hot inland, when the air is dry and the whole countryside is tinder – invoke a deep dread. The wind kicks up memories of smoke-darkened skies; of our neighbour's charcoaled fence posts; of my father coming home from fighting the fires, his eyes reddened, his face blackened with ash; of driving past paddocks where dead sheep lay on the scorched ground, their bodies bloated, their four legs in the air; of red moons rising like swollen avatars of doom.

Everyone knows that we have bushfires in Australia, that it's part of the ecology. And everyone knows that these recent bushfires are different.

On Ash Wednesday in 1983, I was a journalist working for the Melbourne *Herald*, watching the reports come in on the telex machine. That night I went to a party in Port Melbourne, and watched glowing embers fall from the sky in the middle of the city. Forty-seven people died in Victoria, twenty-eight in South Australia.

In 2009, on Black Saturday, I was living in Williamstown. The day was so hot that the leaves of the mirror plant in our concrete backyard turned black. That was the first time I knew that something was different, that this was more than the usual catastrophe. I remember the dread coiling in my gut as I scrolled through the news that day. There was no news coming out of some places. *No news.* I remember watching the death toll rise over the next few days to 173 people: the worst death toll from a bushfire since colonisation.

This New Year's Eve it was different again. Fires that had already been burning for months joined up into megafires. Photos were posted on Twitter that I had never seen before: people fleeing in boats from Mallacoota or houses burning underneath a sky of glowing crimson. At first I thought it was some trick of the camera, but I scrolled through photo after photo after photo, all showing the same glowing red. It looked like the air was on fire.

And then, for months afterwards, there were the photos of Canberra and Sydney, when the sun was a glaring red eye at midday, when the city skylines disappeared in the haze of smoke. The scale was, and remains, impossible to comprehend: the reported deaths of three billion animals, the long-term health effects, the clouds of smoke that drifted over New Zealand and changed the stratosphere.[1] The grief, rising thick in my throat, for everything that has been irrevocably lost.

New Year's is, of course, a purely imaginary division of time. But I think all of us knew, as we woke up in 2020, that this year was going to be different.

Because I'm a writer, I spend most of my day in my home office, witnessing the world through windows. My study looks out through lace curtains onto the front garden. I live in a modest suburb in inner-west Melbourne, where nearly every house, including ours, has roses and some fashion of picket fence.

We came to this house in 2019 – an exhausting move, forced by the sale of our last rental, that came at the end of an exhausting year. At the time, having to move felt like a curse but, ever since I've been here, my primary feeling has been gratitude. Gratitude that we're paying less rent for a house that has no cracks in the walls and a decent bathroom. Gratitude that there's air conditioning. Gratitude that we have a home at all. I keep touching wood superstitiously: we don't deserve our luck. I can't help wondering when it will run out.

Having a garden feels sweet and old-fashioned. When we were house hunting, I was shocked by how many houses were penned in by decking or concrete or astroturf, with no garden of any kind. An agent showed us around a shabby home in Spotswood, with a scrubby, stark back lawn. Nestling against the fence were freshly sawn stumps. 'We got rid of the trees,' the agent remarked. 'They weren't doing anything.'

I think of the earth suffocating and growing sour under all those structures across greater Melbourne, how there's nothing for birds or insects, for all the grubs and bacteria and fungi that turn leaf litter into rich humus. I think of all the trees massacred because they're not 'doing anything'.

It's a weirdly settler-Australian thing, this hatred of trees. When I was a child, we lived on a small farm in western Victoria. Our neighbour, Mr Pearce, had a paddock next to our back paddock in which there was a single, rather beautiful locust tree. One day he cut it down, I suppose because it wasn't 'doing anything'. It was such a shockingly reasonless act: he never bothered to grub out the stump, so he was still forced to plough around it. He just hated that tree being there. I thought he hated it because it was beautiful.

Where do I begin? Did all this start in the 1980s, when Exxon read the reports on greenhouse gases and the company coldly decided to pour millions of dollars into disinformation campaigns to undermine action against climate change? Or should we go back earlier, to the Industrial Revolution? Should we reach back to the 1500s, to the beginning of colonisation, when the superpowers of Europe began to decimate Indigenous cultures and lives and, with them, their knowledge of the environment?

Like everyone else, I perceive the world through windows framed by my background, my history, my gender, my personal inclinations. In the digital age, there are so many different kinds of windows – millions, billions of them, each showing a different, partial view of reality – that orienting yourself within them becomes a challenge.

When we locked down for the pandemic, windows were all we had. I lost some of mine. Theatre and dance have been a huge part of my life for decades – professionally, socially and personally. In a single week in March, with a shocking suddenness, there wasn't any more live performance.

On 13 March, the upcoming Melbourne Comedy Festival – the largest ticketed event in Australia – announced its cancellation. And then the cascade began. Over the following week, I watched as culture across Australia shut down. The pandemic hit live performance – and any activity that requires people to be together in the same space, such as writers' festivals or exhibitions – especially hard.

As with the bushfires, it was difficult to comprehend the scale of the disaster. The Australian website ILostMyGig.net.au asked arts workers whose incomes had suffered from the lockdowns to tally their losses. In the end the website logged more than $340 million of lost income affecting 600,000 people. Theatres lost 90 per cent of their annual revenue. And that was just the immediate effect – the pandemic closures hit an arts sector already cut to the bone after almost a decade of savage cuts.

Now we gather online, watching actors perform live through the small window of a screen, because it's too dangerous for us all to breathe the same air. So begins the voyage into the unknown. Artists have always lived with precarity, and those uncertainties have been amplified into catastrophe. The long-term effect remains incalculable. What shape will our culture be in five years? Ten years?

The last show I saw was on 12 March. It was Caroline Guiela Nguyen's beautiful, epic play *Saigon*, touring from France with her company Les Hommes Approximatifs as part of the Asia TOPA festival in Melbourne. It was a reminder, as I said in my review of it, 'that a mainstage play need not be, as we see too often in our state theatres, a mediocre branch of commercial theatre.'

It struck me then how poverty-stricken our view of main-stage theatre is in Australia. Any work that has ambitions for its vision rather than its box office is largely considered to be the province of independent theatre. When, against all the odds, those artists begin to find an audience, the larger theatres swoop in to pick over and exploit their ideas. When I consider what our under-resourced theatre makers achieve in the teeth of constant precarity and indifference, I am often astonished. Imagine what they could do if that imagination and grit were properly supported? If the so-called 'arts industry' actually valued artists?

It's been commonly remarked that the pandemic has starkly exposed the existing inequities in our societies. It seems that every problem that has rumbled beneath our feet for the past sixty years has hit crisis point, all at once. In the successive shocks of 2020, the gulf between those who have power and agency and those who don't has become bitterly clear. Wounds and conflicts that have long festered in silence and fear are being torn open.

The Black Lives Matter protests that are still galvanising cities around the world are one very visible response to this exposure. But it's happening at granular levels everywhere. Every day I read another story of a life ruined by institutionalised bigotry, of voices silenced – sometimes violently, sometimes covertly – by the foundations of white supremacy that shape our psyches from the moment we are born. Our perceptions and ideas are so stunted by such impossible binaries – white/black, male/female, right/wrong, winners/losers, innocent/guilty – that we are unable to see our world as it is. We are even unable to see ourselves.

Arts and culture are no exception to this reckoning. I believe that art should be one of the solutions but, as we have shaped it now, it's part of the problem. Australian arts institutions are as classist, racist, sexist, homophobic, transphobic and ableist as in any other sector. In fact, they are worse: women arts workers earn 25 per cent less than men, compared with 16 per cent less in the wider workforce. The representation of people of colour in positions of authority is disgraceful.

If art only reproduces the inequities of wider society, what is it for? It's just a rich person's hobby, bling for the apocalypse. Do I want that? I have never wanted that. It's not what I have ever believed art is. I believe that art is a technology for consciousness: it brings connection, it brings recognition. It can allow us to see ourselves and our societies, it helps us to understand how all our truths are complex. It shows us that we are both alone and not alone in our all vulnerable, joyous, contradictory humanity.

I sit in my study, looking out my window at the roses, now brutally pruned for winter. I open news websites on my computer and read of a world riven by fear and violence and injustice. *Where do I begin?*

I think. We must never go back to normal.

Alison Croggon is a writer and editor who lives in Melbourne. Her work includes novels, poetry, plays, opera libretti and criticism.

The great unravelling

JOËLLE GERGIS

If you've ever been around someone who is dying, it may have struck you how strong a person's lifeforce really is. When my dad was gravely ill, an invisible point of no return was gradually crossed, then suddenly, death was in plain sight. We stood back helplessly, knowing that nothing more could be done, that something vital had slipped away. All we could do was watch as life extinguished itself in agonising fits and starts.

As a climate scientist watching the most destructive bushfires in Australian history unfold, I felt the same stomach-turning recognition of witnessing an irreversible loss.

The relentless heat and drought experienced during our nation's hottest and driest year on record saw vast tracts of our remnant native forests go up in smoke. We saw terrified animals fleeing with their fur on fire, their bodies turned to ash. Those that survived faced starvation among the charred remains of their obliterated habitats.

During Australia's Black Summer, more than three billion animals were incinerated or displaced, our beloved bushland burnt to the ground. Our collective places of recharge and contemplation changed in ways that we can barely comprehend. The koala, Australia's most emblematic species, now faces extinction in New South Wales by as early as 2050.

Recovering the diversity and complexity of Australia's unique ecosystems now lies beyond the scale of human lifetimes. What we witnessed was intergenerational damage: a fundamental transformation of our country.

Then, just as the last of the bushfires went out, record-breaking ocean temperatures triggered the third mass bleaching event recorded on the Great Barrier Reef since 2016. This time, the southern reef – spared during the 2016 and 2017 events – finally succumbed to extreme heat. The largest living organism on the planet is dying.

As one of the dozen or so Australian lead authors involved in consolidating the physical science basis for the United Nations' Intergovernmental Panel on Climate Change (IPCC) Sixth Assessment Report, I've gained terrifying insight into the true state of the climate crisis and what lies ahead. There is so much heat already baked into the climate system that a certain level of destruction is now inevitable. What concerns me is that we may have already pushed the planetary system past the point of no return. That we've unleashed a cascade of irreversible changes that have built such momentum that we can only watch as it unfolds.

Australia's horror summer is the clearest signal yet that our planet's climate is rapidly destabilising. It breaks my heart to watch the country I love irrevocably wounded because of our government's denial of the severity of climate change and its refusal to act on the advice of the world's leading scientists.

I mourn all the unique animals, plants and landscapes that are forever altered by the events of our Black Summer. That the Earth as we now know it will soon no longer exist. I grieve for the generations of children who will only ever experience the Great Barrier Reef or our ancient rainforests through photographs or David Attenborough's documentaries. In the future, his films will be like watching grainy archival footage of the Tasmanian tiger: images of a lost world.

As we live through this growing instability, it's becoming harder to maintain a sense of professional detachment from the work that I do. Given that humanity is facing an existential threat of planetary proportions, surely it is rational to react with despair, anger, grief and frustration. To fail to emotionally respond to a level of destruction that will be felt throughout the ages feels like sociopathic disregard for all life on Earth.

To confront this monumental reality and then continue as usual would be like buying into a collective delusion that life as we know it will go on indefinitely, regardless of what we do. The truth is, everything in life has its breaking point. My fear is that the planet's equilibrium has been lost; we are now watching on as the dominoes begin to cascade.

With just 1.1°C of warming, Australia has already experienced unimaginable levels of destruction of its marine and land ecosystems in the space of a single summer. More than 20 per cent of our country's forests burnt in a single bushfire season. Virtually the entire range of the Great Barrier Reef was cooked by one mass bleaching event. But what really worries me is what our Black Summer signals about the conditions that are yet to come. As things stand, the latest research shows that Australia could warm up to 7°C above pre-industrial levels by the end of the century. If we continue along our current path, climate models show an average warming of 4.5°C, with a range of 2.7–6.2°C by 2100. This represents a ruinous over-shooting of the Paris Agreement targets, which aim to stabilise global warming at well below 2°C, to avoid what the UN terms 'dangerous' levels of climate change.

The revised warming projections for Australia will render large parts of our country uninhabitable. The Australian way of life unliveable, as extreme heat and increasingly erratic rainfall establishes itself as the new normal. Researchers who conducted an analysis of the conditions experienced during our Black Summer concluded '... under a scenario where emissions continue to grow, such a year would be average by 2040 and exceptionally cool by 2060.'

It's the type of statement that should jolt our nation's leaders out of their delusional complacency. Soon we will be facing 50°C summer temperatures in our southern capital cities, longer and hotter bushfire seasons, and more punishing droughts. We will be increasingly forced to shelter in our homes as danger-ous heat and oppressive smoke become regular features of the Australian summer. Looking back from this future, the corona-virus lockdown of 2020 will feel like a luxury holiday.

Australia's Black Summer was a terrifying preview of a future that no longer feels impossibly far away. We've experi-enced, firsthand, how unprecedented extremes can play out more abruptly and ferociously than anyone thought possible. Climate disruption is now a part of the lived experience of every Australian.

We are being forced to come to terms with the fact that we are the generation that is likely to witness the destruction of our Earth. We have arrived at a point in human history that I think of as the 'great unravelling'. I never thought I'd live to see the horror of planetary collapse unfolding.

As an Australian on the frontline of the climate crisis, all I can do is try to help people make sense of what the scientific community is observing in real time. I use my writing to send out distress beacons to the wider world, hoping that processing the enormity of our loss through an international lens will help us feel the sting of it. Perhaps, then, we will finally acknowledge the terribly sad reality that we are losing the battle to protect one of the most extraordinary parts of our planet.

I often despair that everything the scientific community is trying to do to help avert disaster is falling on deaf ears. Instead, we hear the federal government announcing policies ensuring the protection of fossil fuel industries, justifying pathetic emission targets that will doom Australia to an apocalyptic nightmare of a future.

The national conversation we urgently needed to have fol-lowing our Black Summer never happened. Our collective trauma was sidelined as a deadly pandemic took hold. Instead of grieving our losses and agreeing on how to implement an urgent plan to safeguard our nation's future, we became pre-occupied by whether we had enough food in the pantry, whether our job or relationship would be intact on the other side of the lockdown. We were forced to consider life and death on an intensely personal level.

When our personal safety is threatened, our capacity to handle the larger existential threat of climate change evaporates. But just because we can't face something doesn't mean it disappears.

As many trauma survivors will tell you, it's often the lack of an adequate response in the aftermath of a traumatic event, rather than the experience itself, that causes the most psychological damage. When there is no acknowledgement of the damage that has been done, no moral consequences for turning a blind eye, it's as if the trauma never happened.

How can we ever re-establish trust in the very institutions that let things get this bad? How do we live with the knowledge that the people who are meant to keep us safe are the very ones allowing the criminal destruction of our planet to continue?

Perhaps part of the answer lies in T. S. Eliot's observation that 'humankind cannot bear very much reality'. To shy away from difficult emotions is a very natural part of the human condition. We are afraid to have the tough conversations that connect us with the darker shades of human emotion.

We are often reluctant to give voice to the painful feelings that accompany a serious loss, like the one we all experienced this summer. We quickly skirt around complex emotions, landing on the safer ground of practical solutions like renewable energy or taking personal action to feel a sense of control in the face of far bleaker realities.

As more psychologists begin to engage with the topic of climate change, they are telling us that being willing to acknowledge our personal and collective grief might be the only way out of the mess we are in. When we are finally willing to accept feelings of intense grief – for ourselves, our planet, our kids' futures – we can use the intensity of our emotional response to propel us into action.

Grief is not something to be pushed away; it is a function of the depth of the attachment we feel for something, be it a loved one or the planet. If we don't allow ourselves to grieve, we stop ourselves from emotionally processing the reality of our loss. It prevents us from having to face the need to adapt to a new, unwelcome reality.

Unfortunately, we live in a culture where we actively avoid talking about hard realities; darker parts of our psyche are considered dysfunctional or intolerable. But trying to be relent-lessly cheerful or stoic in the face of serious loss just buries more authentic emotions that must eventually come up for air.

As scientists, we are often quick to reach for more facts rather than grapple with the complexity of our emotions. We think that the more people *know* about the impacts of climate change, surely the more they will *understand* how urgent our collective response needs to be. But as the long history of humanity's inability to respond to the climate crisis has shown us, processing information purely on an intellectual level simply isn't enough.

It's something Rachel Carson – the American ecologist and author of *Silent Spring*, the seminal book warning the public about the dangerous long-term effects of pesticides – recognised nearly sixty years ago. She wrote: 'It is not half so important to know as to feel . . . once the emotions have been aroused – a sense of the beautiful, the excitement of the new and unknown, a feeling of sympathy, pity, admiration or love – then we wish for knowledge about the object of our emotional response. Once found, it has lasting meaning.' In other words, there is great power and wisdom in our emotional response to our world. Until we are prepared to be moved by the profoundly tragic ways we treat the planet and each other, our behaviour will never change.

On a personal level, I wonder what to do in the face of this awareness. Should I continue to work my guts out, trying to produce new science to help better diagnose what's going on? Do I try to teach a dejected new generation of scientists to help fix the mess humanity has made? How can I reconcile my own sense of despair and exhaustion with the need to stay engaged and be patient with those who don't know any better?

While I hope this will be the summer that changes every-thing, my rational mind understands that governments like ours are willing to sacrifice our planetary life-support system to keep the fossil fuel industry alive for another handful of decades. I am afraid that we don't have the heart or the courage to be moved by what we saw during our Black Summer.

Increasingly I am feeling overwhelmed and unsure about how I can best live my life in the face of the catastrophe that is now upon us. I'm anxious about the enormity of the scale of what needs to be done, afraid of what might be waiting in my inbox. Something inside me feels like it has snapped, as if some essential thread of hope has failed. The knowing that sometimes things can't be saved, that the planet is dying, that we couldn't get it together in time to save the irreplaceable. It feels as though we have reached the point in human history when all the trees in the global common are finally gone, our connection to the wisdom of our ancestors lost forever.

As a climate scientist at this troubled time in human history, my hope is that the life force of our Earth can hang on. That the personal and collective awakening we need to safeguard our planet arrives before even more is lost. That our hearts will lead us back to our shared humanity, strengthening our resolve to save ourselves and our imperilled world.

Joëlle Gergis is an award-winning climate scientist and writer from the Australian National University. Her book *Sunburnt Country: The history and future of climate change in Australia* is available through MUP.

Drawing breath

TOM GRIFFITHS

What was the most shocking event of 2020? Was it awakening on New Year's Day to more news of terror in Australia's southern forests, to the realisation that the future was suddenly here, that this spring and summer of relentless bushfire was a planetary event? Was it the silent transmission of COVID-19, already on the loose and soon to overwhelm the world and change the very fabric of daily life everywhere at once? Or was it the surging race riots and protests around the world, especially across America, where police brutality triggered grief, anger and outrage about the inequality and injustice still faced by black people? Could we even distinguish them from each other, this overlapping sequence of horrors? Fire, plague and racism are always with us, percolating away, periodically erupting, sometimes converging. They came together in the colonisation of Australia when the conquering British brought smallpox, scorned Indigenous rights and fought Aboriginal fire with gunfire. Systemic racism *is* the virus, declared Black Lives Matter protestors.

In mid-June 2020, historian Geoffrey Blainey, writing in *The Australian* in defence of colonial statues, looked back on the first day of the year. His opening gambit was this sentence: 'On New Year's Day, no major economist, no famous medical scientist and no political leader had predicted that this would be a tumultuous year.' Only a defiant climate sceptic could have been so uncurious about the events unfolding that day and so dismissive of expertise. For on New Year's Eve, the Savage Summer had pulsed into frightening ferocity on the NSW South Coast and in East Gippsland. Eight people died that day in the fires. Sleepless tourists and residents faced the dawn of the new year without power, fuel or mobile-phone reception, and some without homes. The day brought evacuations, road closures, panic buying, collective fear and a surge of dire predictions. For months, experts and bush residents had been preparing for a tumultuous fire season and here on the first day of 2020 was a frightening climax. Throughout 2019, fire experts had pleaded with the federal government to hold a bushfire summit to prepare for the dreaded summer, but the prime minister had refused. The crisis could not be acknowledged in case it gave credence to climate action.

As if neglect and omission in the face of the fire threat were not enough, Coalition politicians and their apologists then hastily encouraged lies about the causes of the fires, declaring that they were started by arsonists and that greenies had prevented hazard-reduction burns. Yet these fires were overwhelmingly started by dry lightning in remote terrain, and hazard-reduction burning is constrained by a warming climate.

The effort to stymie sensible policy reform after the fires was as pernicious as the failure to plan in advance of them.

There was barely a moment to breathe between bushfire and COVID. Australians had been in lockdown for months even before the year began, fighting fires that had started at the end of winter, cowering indoors from smoke, heat and ash, and wearing masks on their brief forays outside. People spoke courageously of 'the new normal', but did not yet understand that 'normal' was gone. Just as they finally stepped outside to sniff the clearer autumn air, it was declared dangerous again. Their masks were still in their pockets.

In spite of the connections between these crises, politicians were keen to separate them, as if one blessedly cancelled out the other, not least because the pandemic gave the prime minister a chance to reset after his disastrous summer. Instead of forcing handshakes he was forced to withhold them. For the beleaguered Coalition government, COVID seemed to provide the escape it wanted from climate politics.

Australians were forbidden to talk about the obvious relationship between bushfires and climate, so how will we manage to interrogate the common origins of climate change and the pandemic? The fires and the plague are both symptoms of something momentous that is unfolding on Earth: a concentration and acceleration of the impact of humans on nature. As environmental scientists Inger Andersen and Johan Rockström argued in June 2020, 'COVID-19 is more than an illness. It is a symptom of the ailing health of our planet.'[1]

Or, as American science writer David Quammen succinctly put it, 'We made the coronavirus.'[2] Not in a laboratory but in the scary, runaway experiment humans are conducting with Earth. Historians and scientists predicted the unpleasant surprises. Like most infectious diseases in the history of humanity, COVID-19 spilled over from wild animals to humans and became a pandemic because of ecosystem destruction, biodiversity loss, climate change, pollution, illegal wildlife trade and increased human mobility. 'So when you're done worrying about this outbreak,' warns Quammen, 'worry about the next one. Or do something about the current circumstances.'

Doing something about it means more than finding a vaccine; it means urgently addressing the causes of the climate emergency and the biodiversity crisis. It means understanding how dire the current rupture is in the long-term relationship between humans and nature.

In November 2019, as forest fires worked their way down the eastern seaboard, I walked for a week in the Australian Alps, my annual pilgrimage to the high country. The wild granite tors, the delicate beauty of the snow gums and the exhilarating freedom of the alpine herb fields have always lifted my spirits. In late spring and early summer, this landscape still carries the memory of snow, of a magic, ethereal otherworld I came to know on skis as a child. Slicks of ice remained tucked under crags. It is a place apart, of subtle colours and sharp air, where ranges of cerulean blue cascade in receding waves to the horizon. But this time the mountains had all gone, swallowed by an apocalypse.

From high in the Kosciuszko National Park, I felt like a refugee from the suffering world of the plains, finding solace in the snowgrass and minty alpine forests. I could see nothing of the world below. In every direction I looked down upon a strange, suffocating orange blanket. This was no mystic lake of fog that would evaporate in the morning sunshine; it was something sinister and malevolent, infusing every scarp and canyon with its sickness. There below me, Australia was burning. Could people still be alive down there, in such dense, acrid smoke? Could they breathe? In the mornings, a temperature inversion kept the ugly blanket below me, but each afternoon my eyes started smarting as smoke infiltrated the alpine valleys, turning the sun red. That smoke killed ten times more people than the flames. It was coming for me and I couldn't go any higher.

This experience of looking down on a burning world brought home to me, perhaps more forcibly than facing the flames below, what the future might hold. One way to make sense of this critical tipping point is the idea that we are now living in the Anthropocene, having left behind the relatively stable Holocene epoch, the period since the last ice age. The Anthropocene – the Age of Humans – places us on par with other geophysical forces such as orbital variations, glaciers, volcanoes and asteroid strikes, and recognises our power in changing the planet's atmosphere, oceans, climate, biodiversity, even its stratigraphy. Earth was first jolted into the Anthropocene by the Industrial Revolution in the late eighteenth century, when people began digging up and burning fossil fuels.

But as I gazed down on the smoke, I remembered an alternative name for this era that has been proposed by historian Stephen Pyne. It is the Pyrocene: a Fire Age, comparable to past ice ages. The Pyrocene puts fire at the centre of the human ecological story and contrasts it with ice. Fire is alive and ice is dead. Fire is at the heart of human civilisation, for we are a fire species. Yet we are also, paradoxically, creatures of the ice. We were born in the Pleistocene, a geological epoch that began two-and-a-half million years ago and introduced a series of rhythmic ice ages – or, to be precise, one long ice age punctuated by regular brief interludes of interglacial warmth. The repetitive glaciations of the Pleistocene, which demanded innovation and versatility, promoted the emergence of humanity on Earth.

The Pyrocene is a more radical category than the Anthropocene. The Anthropocene declares the end of the most recent interglacial period. The Pyrocene goes further, by declaring the end of the much longer and older Pleistocene, the whole epoch of ice ages. It announces the end of the Age of Ice, the beginning of the Age of Fire, and the end of what was truly the Age of Humans. Not the beginning of the Age of Humans, as is suggested by the smugly named Anthropocene, but the end. And Australia is on the frontline of the Pyrocene. This is what I pondered as I watched the acrid orange blanket snake up the alpine gullies towards me. Are we witnessing the beginning of the end? Is this what the Pyrocene looks like? Nowhere to go but up, and no up to go to?

The Anthropocene is primarily a geological signature, whereas the Pyrocene is biological; they are both acts of historical imagination that rupture the conventional periods within which we imagine our existence. They ask us to see human history not as something defined by documents or brought into being by the invention of writing, but as a diverse cultural odyssey that is also a biological story – even a geological one. If humans have become so powerful that we can change the condition of the planet's oceans and atmosphere, then we urgently need to think in deeper time, on a scale where we might better understand the environmental rhythms we are so profoundly disturbing.

And yet, at the beginning of Reconciliation Week in Australia, in the midst of the climate and COVID emergencies and as race riots escalated in the United States, the corporate mining giant Rio Tinto detonated 46,000 years of human history at Juukan Gorge in the Pilbara. It was an act of sacrilege committed even as George Floyd's death unleashed avowals around the world that Black Lives Matter. In the age of COVID, people rallying against black deaths in custody donned masks and carried banners inscribed with Floyd's final words: 'I can't breathe.'

Several years before Juukan Gorge was destroyed, archaeologists found a 4000-year-old belt made of plaited hair in one of its rock shelters. Its DNA was associated with today's Puutu Kunti Kurrama and Pinikura traditional owners. Chris Salisbury, Rio Tinto's CEO of Iron Ore (who resigned come September), apologised for 'the distress' caused by the destruction of the site but not for the act itself, which he defended. Here is staggering evidence of Australia's continuing inability to empathise or identify with the peoples who discovered this continent and who today are still fighting for recognition, justice, respect and equality before the law. It is confirmation that in the twenty-first century our country remains a colony, still unable to accept (as the Uluru Statement puts it) that 'this ancient sovereignty can shine through as a fuller expression of Australia's nationhood'. The deep environmental and cultural inheritance of this continent, with all the wisdom and perspective it might offer about living in this place, about survival, species and cultural burning, about fires, plagues and rising seas, is not yet important enough to Australians. When will it be, if not now? The insidious smoke is coming.

Tom Griffiths is Emeritus Professor of History at the Australian National University. His books and essays have won prizes in literature, history, science, politics and journalism. His most recent book is *The Art of Time Travel*.

Dead water

SOPHIE CUNNINGHAM

One hundred and fifty years ago, Ludwig Becker, a member of the Burke and Wills expedition, did a sketch of the Menindee pub, which sits between the Darling River and Menindee Lakes. At that time, Menindee wasn't considered big enough to be called a town, but these days it needs a school for more than 100 pupils. Three-quarters of these children are descendants of the Barkindji and Nyampa people, who have lived, hunted and passed down their oral histories on the banks of the Darling for more than 40,000 years.

When Burke was camped at Menindee he met with William Wright, a local station manager. Wright was charged with leading a small group to transport supplies to the camp at Cooper Creek. His contingent was joined by Becker. En route to Cooper Creek, Wright's group pitched tents by the Koorliatto Waterhole on the Bulloo River. They were visited by Mr Shirt, a Bandjigali or Karenggapa Murri man whose portrait was also painted for posterity by Becker. Mr Shirt, a 'born diplomat', tried to explain the problem the explorers were causing: 'the area belonged to his tribe. Soon they were coming here to celebrate a feast . . . neighbouring tribes were already coming to drive us away.'[1] Not long after that conversation, Wright shot Mr Shirt.

The theft of water in the Murray–Darling Basin has a long history and it began when Burke and Wills walked from Royal Park to Moonee Ponds, then another 750 kilometres to Menindee, then north again, with little clue as to what they were doing. Seven white men died on the expedition (including Ludwig Becker). Mr Shirt died. Twelve Ngawun men were killed by Frederick Walker, a notorious former Native Police Officer who led one of the expeditions that searched for the remains of the Burke and Wills party in 1861.

The Burke and Wills expedition was the first in Australia to use camels to carry supplies. The descendants of some of these camels still live in outback Australia. The problem is this: a herd of thirsty camels can drink a waterhole dry very quickly. When these early explorers arrived and set up camp, they took precious resources – water, fish – from the Nations whose land they were on. While welcomed for short stays, they were considered a pest when they set up camp for long stints. Following European occupation, the Nations living through the Murray–Darling Basin were steadily dispossessed as the land and waters were exploited for agriculture.

Fast forward to February 2020. Torrential rain fell along the east coast of Australia – more than 350 mm in some places, delivering the highest February daily rainfall on record. On 10 February, the NSW Rural Fire Service tweeted, 'This is the most positive news we've had in some time. The recent rainfall has assisted firefighters to put over 30 fires out since Friday. Some of these blazes have been burning for weeks and even months.'

Floods are destructive, but they used to mean good things as well. They gave life to river systems and wetlands, they flushed out sediments and salt. Rivers met each other in a rush and their mouths opened to the sea. But as droughts grow longer, rain, when it does fall, causes more damage. As temperatures rise, the air holds more moisture, which increases the likelihood of intense rainstorms. When heavy rain falls onto eroded and parched soil it doesn't seep into the ground so much as skid and skim over the top of it.

Insurance claims were filed from south-east Queensland down the New South Wales coastline. Damage claims were also reported several hundred kilometres inland. Floods are estimated to be the most costly natural disaster in Australia, and the Insurance Council of Australia immediately declared the floods in February this year a catastrophe, the sixth catastrophe they'd declared in a *five-month period*. The time we are living through has been described as an Era of Disasters, a time 'when emergency services will likely be stretched, community resilience undermined, and economic costs and loss of life increased.'[2]

It can take a while for floodwater to reach water catchments, however by mid-March, waters were expected in Menindee for the first time in three years. This was cause for celebration. Menindee had been in survival mode for months, if not years.

Several hundred Murray cod and golden perch were swimming around in an aerated weir pool in the hope that they too, could survive the drying of the Darling River and Menindee Lakes. Menindee man Graeme McCrabb was one of a number of locals who had rescued fish throughout 2019. This meant that 12 March 2020 felt like a good day for a change. McCrabb – indeed, the entire town – was relieved by the arrival of the water, and the life that swam and flew along with it.[3] This water signified the possibility that the Murray–Darling river system might become connected from end to end after many years of being strangled. And indeed, the rivers did join, and they did meet the sea. Just.

But welcome as the water was, there were also concerns. During droughts, organic matter, like leaves, builds up on floodplains and the banks of rivers. When substantial rain finally occurs and water flows over riverbanks onto the floodplain, it collects this debris and dumps it in the river, causing oxygen levels in the water to drop significantly. This kind of flooding is described as 'black water'. It can cause fish to die, a particularly unwelcome possibility given that up to one million fish had *already* died along a 40-kilometre stretch of the Darling River in far west New South Wales during the previous summer. This is why McCrabb and others had been trying to get fish out of the river and into the weir.[4]

WaterNSW forecast that up to 285 gigalitres might reach Menindee after February's rain but it takes 1700 gigalitres of water to fill Menindee's waterways.[5] A look at historical data from Wilcannia River, the stretch of the river that leads into Menindee Lakes, tells you that the highest volume of water to flow into Menindee in a single day was 68,415 megalitres. That was in 1976, just a few years after records began. The strongest flow in 2020 occurred on 19 March, the day before the waters reached Menindee and was a mere 13,080 mega-litres. Not enough to compensate for the fact that the previous year had been the worst stream flow on record. Not enough to compensate McCrabb and the many others who'd stood above Menindee's Weir 32, watching in horror as tens of thousands of golden perch flapped and strained and 'tens of thousands of little bony bream' lay dead.

At the time, those (multiple) fish kills were described by the NSW government as 'distressing' and unavoidable. 'We cannot control the weather,' Premier Berejiklian said.

They were avoidable. And, while we may not be in control of it, the weather is on us as well.

The Murray–Darling Basin covers nearly a seventh of the Australian continent and touches on the traditional land and waters of thirty-four Indigenous Nations. In 2015, the Federal Court recognised the rights of the Barkindji peoples to 128,000 square kilometres of land, which included both sides of almost the entire length of the Darling River. At the time of the fish kills, Barkindji elder William Badger Bates remem-bered that one of the arguments used to harvest water from the Menindee Lakes was that if it was left, the water would simply evaporate. He called bullshit. 'Don't anyone try to tell me or my people or the rest of the people that these lakes evaporate . . . we have known them for thousands of years before,' Bates said.[6]

What is true is that the lakes are in a semi-arid area, they are shallow and have a large surface area. In most years the equivalent of one Sydney Harbour of water turns to vapour there.[7] But there are many ways for water to evaporate. These include irrigation, fraudulent transactions and water theft. On 3 September the Wentworth Group of Concerned Scientists released a report that states that 20 per cent of the water that the government estimated should be in the Murray–Darling Basin system is simply 'missing'. On average that is 320 billion litres unaccounted for every year since 2012. This is despite the government having spent $7 billion to improve the health of the river system.[8] The entire basin has been starved of water and suffocated by a system of bureaucracy that is confusing to the point of being impenetrable, and, as a consequence, easy to corrupt. To quote journalist Margaret Simons, 'It's impossible for ordinary citizens to find out who owns water, or who has made a trade.'[9]

After the fish kills, there were various emergency summits. 'No one expects the river to run every year,' said Professor Craig Moritz, who chaired an Australian Academy of Science panel, 'but they have cut the water so hard, the river is dying.' Badger Bates would have agreed with this assessment, but, despite being a traditional owner, only found out about these summits after they'd been held. After this 'oversight', an amendment to the *Water Act* was finally passed in October 2019 that established a position for an Indigenous person on the Murray–Darling Basin Authority board.

The Murray–Darling Basin Plan sets limits on how much water can be taken from the Basin for irrigation, drinking water, industry or for other purposes. But, as Professor Jamie Pittock of the ANU's Fenner School argues, 'The basin plan was written for a static environment and is not adapting to climate or other changes.' Pittock and others have also pointed out that some irrigators are more successful at exploiting the cap system, which has been described as many things, perhaps most evocatively as 'leaky'.

Pittock, who is a member of the Wentworth Group of Concerned Scientists, describes Menindee as the standout example of mismanagement in the entire Murray–Darling area. Its inadequacy in the face of a rapidly changing, increasingly challenging environment keeps him awake at night. 'Our government fails to recognise, monitor and manage these risks,' he told me. 'By the late 1990s water storages were in operation that could hold around three times the river system's average natural flow to the sea and the diversion of water for irrigation resulted in severe environmental impacts.'[10]

Pittock isn't the only one losing sleep – most people who are living in the Murray–Darling Basin are enduring a nightmare. Since the late 1990s, there has been a decline in rainfall of around 11 per cent over the cooler months in the southeast of Australia, and projections make clear that this decline will continue.[11] At the same time, as climate change has begun delivering increasingly heavy blows, thirsty crops are the equivalent of Burke and Wills's camels, guzzling a limited water supply.

Scientists who work in the Murray–Darling Basin have resorted to medical metaphors to describe what has been happening to this ecosystem.[12] Wetland scientist Richard

Kingsford likened the Macquarie marshes to 'a very ill patient who has been given just enough care to get out of the ICU'.[13]

People of the Barkindji nation are blunter still. At the time of the fish kills, Lilliana Bennett recalled her grandmother talking about going down the riverbank to fish and hunt for goanna. The Darling River was an important place for her family.

'It's a place they go to relax, to tell stories,' she told SBS News. 'For me, it's been really devastating, I mean, we went down and camped by the river where there's still a bit of water around and it just doesn't have the same feeling, it's dead water.'[14]

Sophie Cunningham is the author of six books, including *City of Trees*, *Melbourne* and *Warning: The story of Cyclone Tracy*. She is a former publisher and editor, and is currently an Adjunct Professor at RMIT University's Non/fiction Lab.

A matter of urgency

TIM FLANNERY

I was in Melbourne in late January, watching as more and more people donned face masks to protect themselves against the bushfire smoke that had thickened the air for weeks and that was causing hundreds of deaths. Turning on the news, I was surprised to see footage of crowds in China similarly masked, but for a very different reason. Hundreds were then dying in Wuhan, Hubei Province, from a novel virus.

When I asked Australia's chief medical officer about the virus that same week, I could see the concern in his eyes. But my attention was largely on the fires. They were unlike anything experienced on the continent previously, and climate scientists were beginning to piece together the link with climate change. What few knew back then was that three catastrophes would strike Australia in quick succession: the unprecedented, climate-fuelled megafires that were extinguished in February by damaging, climate-influenced floods. Then, in

March, the COVID-19 pandemic that began to spread across Australia.

These three catastrophes are proof that things that travel invisibly through the great aerial ocean that is our atmosphere are a particular danger to our complex, global civilisation. The CO_2 molecule that accumulates imperceptibly as we burn fossil fuels causes an increase in average global temperature, which triggered the profoundly disruptive droughts, floods and fires that plagued Australia over the past year. But the coronavirus also travels unseen through the great aerial ocean, insinuating itself in lung after lung, killing person after person, until it threatens our health system, economy and society.

There are many differences between climate change and the COVID-19 pandemic; but from the perspective of prevention, there are also many similarities. Perhaps the most important is that both have 'incubation periods' during which the problem grows, undetected, except by the experts. Throughout this period, things can seem relatively normal, but unless a sense of urgency leads to decisive action at this time, catastrophe becomes inevitable.

The actions required to contain both a pandemic and climate change are also broadly similar, and involve three steps. The first and most urgent is to stop the threat from growing. For COVID-19, that involved introducing social distancing, closing schools, and halting entire industries. For climate change it means dramatically cutting the use of fossil fuels. The second step involves ensuring that we can save as many of the stricken as possible. For COVID-19, that meant preparing emergency wards and other treatment facilities. For climate change it means instituting measures to deal with a sweeping variety of issues, including future megafires, the threat to the Great Barrier Reef, and vulnerable coasts. The third step involves finding a permanent fix. For COVID-19, that means the development of a vaccine, while for climate change it involves removing the excess CO_2 from the atmosphere.

Many Australians have been astonished at the contrast between the federal government's responses to the pandemic and to the climate threat. It was missing in action for much of the climate-related megafire and flood crisis, but in the face of a pandemic it acted swiftly. A real sense of urgency, prompted by scientific advice, was evident when Australia cancelled flights from China in February (well before most other nations had acted), and when it labelled the COVID-19 threat a global pandemic twelve days before WHO announced that it was upgrading the threat to that status. But bigger things were to come. In the middle of March, by which time the number of infections in Australia was doubling every four days, the Morrison government locked the nation down, dealing a devastating blow to the economy, but saving thousands of lives.

Overall, Australia has mounted one of the most effective responses to the virus of any country. Yet on climate change it remains unalarmed and unmotivated. This may prove catastrophic, for the climate emergency is now entering a crucial phase. In COVID-19 terms, we are in mid-March – the last possible moment for emergency action. That's because the concentration of greenhouse gases in the atmosphere is now so great that to delay even for a few more years risks triggering Earth's tipping points. And if that happens, there will be no way back.

Researchers have identified fifteen tipping points for Earth's climate system. They involve events like the melting of glaciers and ice caps, the destruction of the Amazon's forests, and the altering of ocean currents. Trigger any of them, and a cascade of consequences is unleashed that will lead to out-of-control planetary heating. Trigger the tipping points, and almost everything about Earth will change, including biodiversity, the coasts, our food and water security, and our health.

The urgency of our situation was recently underlined by Australia's most eminent climate scientist, Professor Will Steffen. In an interview with *Voice of Action* on 5 June, Steffen stated that 'we are already deep into the trajectory towards collapse' of our civilisation, because nine of the fifteen known global tipping points have already been activated.[1] This is the equivalent of the nation's chief medical officer telling the prime minister in March that he must act today if he wishes to contain the pandemic.

To date, government action on climate change has been so tardy that very soon, enough CO_2 will be in the atmosphere to make it impossible to achieve the lower Paris Agreement target of keeping warming below 1.5°C. And, within a few years of that, there will be enough to make the higher target of 2°C unobtainable. Time is now so short that we cannot wait for the next Australian election for action. It is the Morrison government that must act decisively if Australia is to do its part in averting this looming disaster. Despite the obvious impediments and appalling track record of some Coalition governments, looking back on the events of early 2020, I have hope that the Morrison government can lead Australia out of danger.

Sometimes, it takes a terrible disaster to alert people, and Australia's megafires may well be the moment when we as a nation awoke to how exquisitely vulnerable our country is to the effects of climate change. Historically, in a bad bushfire year, around 2 per cent of Australia's forests will be burnt. But in the summer of 2019–20, more than 21 per cent of the country's forests was aflame. That's a tenfold increase, and it's the kind of step change that we're increasingly seeing as our climate system begins to destabilise.

The link between the megafires and climate change is clear. South-eastern Australia has been getting hotter and drier for decades, and 2019 was Australia's hottest and driest year on record, with 2018 being equally dry over south-eastern Australia, and almost as hot. Fires are profoundly influenced by temperature and dryness (which is why they occur in summer rather than winter), and the long, hot, dry spell of 2018–19 set Australia's forests up for burning. Fire chiefs had been warning of the danger for months, but the prime minister had refused a meeting to discuss the growing emergency. He even went on holiday as the fires began to peak. By the time widespread flooding extinguished the fires in February 2020, thirty-four people had died in the flames, nearly 3000 houses had been destroyed, and entire regional economies were in tatters.

Beginning with the United Kingdom in May 2019, one nation after another has proclaimed a climate emergency. And they are acting strongly to deal with that emergency. By mid-June 2020, the UK (the country where industrial coal-burning started) had gone two months without burning coal. But Australia has neither declared a climate emergency nor acted decisively. Despite our abundant sunlight and wind resources, we are still 60 per cent dependent on coal for our electricity needs.

There could not be a clearer case of the dangers of inaction in the face of the climate emergency than Australia's megafires. The extent of the reasons why the federal government is not treating the climate emergency as it did the health emergency is probably known only to Morrison's cabinet. But a few factors are evident to all.

That Australia is the world's largest exporter of gas and coal, two of the three fossil fuels (along with oil) that are causing climate change, is clearly fundamental. Too many people, including some politicians, are doing far too well from the trade in fossil fuels to want to stymie it, regardless of the impact on the rest of us. With coal in global decline and few oil resources, gas is the healthiest sector of Australia's fossil fuel industry, and it is the gas sector that the Morrison government is focusing on to lead the post-COVID recovery.

But if economic opportunity were the only driver of climate denialism, it could be countered by creating opportunity elsewhere, and to some extent this is happening. With enormous potential to be found in green hydrogen and the renewables sector, some bright young people are leaving the fossil fuel industry and staking their futures on the new, clean economy. What is holding back progress most strongly is the AUD $80 billion that corporations have invested in domestic gas infrastructure. Acting on the climate emergency would mean that these corporations will face huge losses. In ignoring the climate scientists and investing so heavily in gas they have made a bad economic bet, but are unwilling to face the consequences.

Interwoven with self-interest, the Morrison government suffers from a thick strand of climate denialism that feeds on tribalism and wilful ignorance. Ex-prime minister Malcolm Turnbull believes that the Coalition continues to struggle with climate denialism. But there has been a shift, at least in terms of rhetoric, since the megafires. It's been a while since we heard climate lunacy from the mouth of Craig Kelly, and two of the most adamant denialists in the National Party, Barnaby Joyce and Matt Canavan, are now languishing on the backbench.

I think that as the full consequences of the megafires begin to be understood, climate denialism will become more and more difficult to sustain. The megafires, like the pandemic, took many lives. By mid-June COVID-19 had killed 102 Australians, but smoke inhalation from the megafires alone killed 443. Nor is the economic impact of the megafires insignificant: estimates put the full cost between AUD $100 and 200 billion. Because the damage is concentrated in certain regions, those communities will suffer for years as they strive to rebuild. Finally, there are likely to be more megafires in the future. If we had not already added so much CO_2 to the atmosphere, we could expect conditions as hot and dry as those of 2019 to occur around once every 400 years. But, due to the levels greenhouse gases had reached by 2013, that probability has increased to once every eight years.

Will the Morrison government act in time? There is one important difference between the pandemic and the climate emergency that may hinder prompt climate action. Pandemics grow quickly: one week there might only be a scatter of cases, but within a fortnight, without strong action, there could be thousands. By comparison, the climate emergency is slow-moving. The fate of Malcolm Turnbull warns that those struggling against self-interest and climate denialism have a difficult job ahead of them.

One cause for optimism, however, is the fact that the mega-fires and the pandemic have exposed some of the lies told to frustrate action on climate change. That it would be 'economy wrecking' to take action in the face of the climate emergency is one. Australian electors now understand that their government can do extraordinary things to protect them.

One thing we could all do right now to help is to challenge the denialists. Before the COVID restrictions, hundreds of people attended a meeting in Sutherland Shire aimed at ousting their local member, Craig Kelly, in order to replace him with a representative who understands the need for climate action. And at the 2019 election, denialist-in-chief Tony Abbott was defeated by an independent, Zali Steggall. Were the denialists visibly challenged everywhere, their grasp of power within the Coalition would slip even before the next election.

Tragically, the news from the climate scientists is getting worse and worse. Increasingly, many experts are viewing 2021, and specifically the United Nations Climate Change Conference to be held in Glasgow late that year, as humanity's last chance to avoid an environmental apocalypse. If there was a moment of true emergency in the fight to preserve our climate, it is now.

Tim Flannery is a paleontologist, explorer and conservationist, a leading writer on climate change, and the 2007 Australian of the Year. His books include the award-winning international bestseller *The Weather Makers*, *Here on Earth* and *Atmosphere of Hope*. He is currently chief councillor of the Climate Council.

Don't blink

JANE RAWSON

I am descended from people who factor a flat tire into a drive to the airport. I own a personal, portable water filter, just in case. I am someone who patrols her boundaries. I am a list-writer, a timetable-checker.

The overarching project of my life has been making myself safe. No alarms; no surprises. It has become legend in my family that, at age eleven, I ruined a holiday by demanding we move out of our accommodation at the foot of what everyone told me was a dormant volcano, because I thought it was too dangerous. (The volcano did erupt, on my thirty-fifth birthday.)

Nothing had changed by age forty-four, when I published a personal guide to surviving climate change, which was essentially a list of everything I was afraid of and all the ways I planned to stop those things happening to me. By age forty-nine my plans had come to fruition. I had left inner-city Melbourne and moved to the Huon Valley in the south of Tasmania.

I'm not the only one who has thought it worthwhile making huge changes to their life in an effort to stay safe. There are the awful white-supremacist preppers, of course, and the billionaire tech magnates with their horrible luxury New Zealand bunkers. Self-sufficiency guru Michael Mobbs caused a ruckus in 2019 when he announced he was selling up his Sydney home to escape the coming societal collapse; he planned to shift to Bermagui, which he thought would be safer. (On 23 January 2020, residents of Bermagui were told the Badja Road fire was heading in their direction and it was too late to leave; to seek shelter as the fire approaches; to protect themselves from the heat of the fire.)

Since Greta Thunberg started making headlines, since the IPCC declared we have only twelve years left to get our act together, since the UN's biodiversity body warned in 2019 of imminent ecosystem collapse, people of run-of-the-mill, middle-class privilege, friends and relatives of mine, have been quietly approaching me, asking, 'Where will I be safe? How can I keep my children safe?'

When my co-author and I wrote our handbook, we tried to answer this question. The answer was: nowhere. There is no *where* that will make you safe, there is only a *when*: when you become rich enough to build your children a bunker village with its own food and water and oxygen; even less probably when we decide to redistribute society's benefits so that being rich is not a pre-condition for being safe.

Until either of those *when*s happen, we suggested, you need to change your definition of 'safe'. Stop looking for places to hide and barriers to put up. Build stronger relationships with the people around you so you will be there for one another when difficulties arise. Invest less of your time accruing material goods (like personal water filters) that you could lose in a fire or flood, and more of your time organising for social change. Stay light on your feet, valuing people and experiences more than you value property, and look for a life where those things are nurturing and exciting.

So much for all that: in my struggle for safety I abandoned the network I had in Melbourne, moved to Tasmania on 16 January 2019, and bought myself some property just east of Huonville. (On 15 January, lightning struck 2400 times across south-western Tasmania, igniting several fires including the Riveaux Road fire, which threatened the Huon Valley for the next three months.)

During the 2019–20 mainland bushfires, when Tasmania was cool and damp, I was safe. During the coronavirus pandemic, isolated in my cottage at the end of a dirt road on a sparsely populated island at the bottom of the world, I was safe. (By late April 2020, an outbreak in the north-west of the state meant Tasmania had the highest number of coronavirus infections per capita of anywhere in Australia.)

Sometime in early April a friend in Melbourne sent me an email. 'You must be feeling pretty pleased with yourself. The fires, now this?!'

I was feeling pretty pleased with myself. I realised that all that work I'd done – the project of a lifetime – had finally paid off.

And then I sat on a rock all by myself and had a little cry because it felt terrible.

We all know, deep down, that we can never be safe. Don't we? Searching for safety is a panicked thrashing around that drags you deeper and deeper into the quicksand. For a while in my twenties I went to the doctor once a week – four different doctors, so, as far as they each knew, I went to the doctor once a month – just to check whether I was safe from dying. They would say yes, and I'd have a day of relaxing, and then I'd think, well, yesterday I was safe, but how about now? What I wanted was to stay in the doctor's office, hooked up to a (so far non-existent) machine that would feed out constant information about my health, alerting me instantly if I was in any kind of danger, for the rest of my life.

You can only ever know if you are safe in retrospect, once the you that was safe is gone forever without ever knowing how safe she was. You can wake up in the morning and know you didn't die in the night, but you can never go to sleep because who knows what the future holds? Where can I move to that is safe? Let me know if you figure out how to move to the past.

Oh, the night. I would also ask my housemate to check on me now and again to see if I had died or was near death while sleeping, because perhaps the only way to relax your vigilance is to hand the burden to somebody else; the trick, then, is not being too vigilant about how well they're carrying it. (She never checked whether I had died.) When I was much younger, I assumed the government, grown-ups and some other amorphous forces were carrying the burden of keeping me safe from existential threats. Because that's their job, isn't it? A child raised in the '70s and '80s – me – could sanely go through their early life assuming that the government was concerned with their health, their financial wellbeing, even the future of the earth they relied on. (By 1985 the world had stockpiled 61,662 nuclear warheads.)

But now? Watching the disintegration of the United States and Brazil, knowing their rampant obsession with individual freedom is no different to here, except a little more heavy-handed with the satire, now I know – and we all do – that each of us must bear our own burden. Nobody else is watching out for us. We cannot afford to relax.

This state of constant vigilance where threat is ever-present, just over there, just outside our line of sight (don't blink), it doesn't leave a lot of time for anything else. It eats everything. You drag yourself panting into a glade in the sunlight of safety and instead of feeling joy, calm, peace, you find the forest is full of ticks (every last one of which is carrying Lyme disease).

This is me, on a rock, having a cry because sure, I'm safe right now, but what is the point of being safe if everyone else is drowning and burning and starving and all the things you love are desiccating in the ever-hotter, ever-drier atmosphere? In a world like this world, safety means isolation and loneliness. It's a jerk act to smile when everyone else is weeping in pain. There are no moments of spontaneous wonder in a bunker.

Where can you be safe? Where can your children be safe? You can't, they can't, stop asking that question. Is there something better to be than safe? Well, I don't know. But maybe it's better to be brave.

I have never managed to be particularly brave. I've always been able to come up with a rationale for slipping back into my comfortable life before things get really bad; my lucky life has always let me. But wouldn't it be something to behave courageously? To see the threat and step forward to meet it?

In 1917, the Australian poet Lesbia Harford wrote:

'Today is rebels' day. Let all of us
Take courage to fight on until we're done —
Fight though we may not live to see the hour
The Revolution's splendidly begun.'

'Politics is hard,' write Kai Heron and Jodi Dean in their article 'Revolution or Ruin' for *e-flux*, 'because it asks us to take and wield power, to be disciplined, focused, and clear-eyed . . . it asks us to choose sides, to name our comrades and our enemies.'[1]

I think about that when I see you all — on the news, on Twitter — out on the streets angrily demanding justice, polarising opinion, while I'm here at the end of my pleasant country road. There is a kind of climate action that is about whittling away your own impact, reducing your footprint, negating yourself. What am I trying to make myself safe from? Bushfires? Floods? Criticism? Judgement? Small and quiet, squeaky-clean, a tiny target: 'Leave me alone,' I beg the world and its terrors, and that is exactly what it does.

Jane Rawson is an author and environmentalist. She has published four books. Her most recent, *From the Wreck*, won the Aurealis Award for best science fiction novel.

Our daily bread

GABRIELLE CHAN

If Australia is known for anything, it is farming food in a brittle environment. We have built many of our national myths and legends around our capacity to produce primarily wheat, sheep, beef and dairy in the midst of bushfires, drought and flood. We also have a healthy food export infrastructure, and are fond of saying we feed twice as many people abroad than we do at home.

Yet 2020 has given us a kick in the arse. The year began with a combination of drought in the eastern states, followed by catastrophic bushfires. Losses of life, biodiversity and property were followed by reports of food shortages, as supermarkets had to close or severely limit their hours due to power restrictions and inability to receive supply. The floods that came after the fires further complicated access to basic needs for communities who were only just starting to stagger to their feet after the bushfires had burnt their landscapes bare. Heavy rain washed away topsoil and the power was cut again. And, just as communities were setting out on the long road to recovery, the pandemic closed in, sending people into isolation of an uncertain duration. These colliding crises have revealed a hole in the system that takes pride of place in our national identity: our ability to provide food to market.

You couldn't miss the weeks of empty supermarket shelves, stripped of staple items like pasta, rice and flour. But unless you were paying attention, you may have missed the short, sharp political storm over whether Australia's food security was in danger. Southern irrigators in the Murray–Darling Basin used the buying panic to write to the federal government, urging the release of water as rice stocks fell. Nine's *60 Minutes* covered the story, forcing the agriculture minister David Littleproud to try to calm shoppers. The National Farmers Federation quickly rolled out a campaign to remind Australians, 'Don't panic, we've got your back.' The Australian Bureau of Agricultural and Resource Economics and Sciences (ABARES) called any food security fears 'misplaced'.

ABARES is right in the sense that most of us won't starve. But the scale of our food producers' operations is hollowing out to either very big or very little, with the middle – family-owned farms that employ few or no staff – falling by the wayside. The statistics back that up. According to ABARES, 'High-revenue farms now account for one fifth of the broadacre population but two thirds of land, income and output.'[1] The mid-sized, multigenerational family farm of the Australian imagination is turning into a fantasy: our rural communities are changing as fast as the climate.

Given these trends, an interesting food subculture has emerged, one fed and watered by the disruption of 2020. In this discombobulating year, the thing people returned to was food. We grew it, we prepared it, we baked it, we ate it with family or alone. As if waking from a long sleep, we got back to the basics. Local food-supply chains were celebrated. Local bakers, greengrocers, community-supported agriculture, community gardeners, millers and farm-to-table meat producers and suppliers have mostly kicked on throughout COVID, proving the value of being close to your customers and relying on local people for your delivery chain.

Many small outfits have told me that their sales have gone through the roof in recent months. Inquiries for regular supplies from small food producers, whether it's meat, vegetable boxes, locally milled and baked bread, or milk have sometimes outstripped capacity. This is particularly true in rural communities, where supplies have usually travelled some distance, even in areas surrounded by farmland. Supply chains often take circuitous routes. For example, stone fruit grown around me in the NSW south-west slopes can be delivered four hours away into city markets, bought by a supermarket chain then delivered straight back to the same area, that much older and more expensive.

Nature's role in the production of food is mostly set. Human intervention is not. I was interested to see that a drive into these bushfire-affected areas revealed green shoots – not just in the landscape but in the people and their food systems as well. People in the tiny towns and foothills of the Upper Murray, trying to recover from losing their houses and town infrastructure, were making it up as they went along. This is the type of growth that comes after devastation and confusion. It is the perfect imperfection of responding to need. It is completely accidental in a way, a combination of fires and pandemic, mixed with personalities who emerge and rise to the occasion. It starts with the seed of an idea, a seed that sprouts as the interest grows and then flourishes with time, attention, money and goodwill.

A few years ago, Josh Collings and his partner Kate Crowley decided they could not afford to buy a house close to a city with kids so they drew a radius of where they wanted to live and set their search filters at $150,000. A little cottage in Cudgewa near the Murray River came up and they moved in, had a son and started renovating. They grew vegetables and Josh ran an art gallery in nearby Corryong called Show and Tell. As Josh got into the garden, he and fellow gardener Jacqui Beaumont established a community food swap, which drew in gardeners from around the Upper Murray. They talked about the idea of turning it into a market garden.

When the fires approached, Josh and his family took the official advice and moved to a shelter. When he returned to the cottage, it was gone: just a tangle of iron and a shell of a burnt-out bus. The only place on the property that was not cinder-black was a square of green zucchinis in the vegetable patch. As they drove around town to survey the damage, they noticed it was a common theme. Green vegetable gardens next to burnt-out buildings.

Roadblocks ringed the town. Dried foods and staples were the only supplies coming in and out. Prices had doubled and the supermarket would only take cash. Josh and Kate bought a bag of food with change scrounged from their wallets and car. It came to just a dollar more than they had. Something had to go back. It was another turning point.

'I had this dream,' he says. 'What was going to be my tiny garden became one-acre gardens in the Upper Murray to create resilience and make sure we always have fresh food. They don't burn because they are so heavy in water and they could produce income for communities who want to be involved.'

Josh started raising money for the recovery two days after their house burnt down. He set up a Go Fund Me page, which quickly raised $190,000 for much-needed generators, food, masks, fencing, tools and fuel. Along with friends Ben Gilbert and Tristan Pierce, Josh delivered these goods to people in need. Some of the money went to keeping Cudgewa's spirits high in the weeks immediately after the fires. Pubs were stocked with free beer, which went nicely with a donated pallet of ribs and other produce. Bands offered to play for the town and this kept people smiling as they gathered to swap stories. Josh made contact with other bushfire-hit towns like Tumbarumba, Mallacoota and Cobargo. He also asked Scott Pape, author of *The Barefoot Investor*, to speak to the locals. Scott encouraged him to keep going, and featured Josh's work in a documentary for Foxtel. Pete Williams of Deloitte offered his advice on recovery and has been working with the community since January. Myriad other organisations also offered time, skills or money.

Then COVID hit. Food supply contracted again. People retreated from the town centre. The fire-recovery hubs that had sprung up fell away. All levels of government stopped their recovery services in fear of spreading the disease.

Josh's usual work shooting video and audio had dried up, so he dedicated his time to the local recovery effort. While the supermarket ensured people were fed, he kept returning to that seed of an idea about a community garden. He had friends who were into permaculture and he started researching. He created the Acres and Acres Co-op: a mash-up of a community garden and a market garden, in which the profit made from vegetables sold pays its workers, with the rest split between the co-operative and community projects.

He teamed up with fellow gardeners Pam Noonan, Jacqui Beaumont, Shelly Neale, Dee McDonald, Lysander Tyrell and many more. The wider community has provided all of the funding so far to buy tools and equipment. The first garden sits on a piece of land behind Josh's art gallery in Corryong, with half of that space gifted to Acres and Acres by an older Corryong resident via a pledge written on the back of a napkin. A load of rich black soil has been delivered, ready to plant with vegetables.

Rather than slowing down during COVID, the team is expanding into other towns in case of food chain collapse.

'The idea is a garden in different towns,' says Josh. 'Cudgewa, Thowlga, Biggera, and so forth, managed by the Acres and Acres Co-op, using the shared tool library. We have fresh food, food resilience, access to jobs if [people] want them and community money for whatever they decide they need, without the painful process of grants. It would be a continuous injection of funds and provide a farm gate for tourism as well as supplying the local supermarket. I want to get all the right people connected to keep the craziness going.

'There has been so much goodwill so far. I'm not exhausted, I get energised by disruption. It has to be done. If we don't do it now, it won't happen. I love to see all these little businesses grow and connect and a few other businesses are forming alongside Acres and Acres, and that is the idea.'

The value of a community building project is not just in connecting people. We are a nation built on bread and steak; only around four per cent of our farms grow vegetables. Which is lucky, because only about seven per cent of us (and five per cent of our kids) eat the daily recommended serves of vegetables, and that proportion is dropping.[2] Similarly, the number of veggie farmers has been declining, with a third of them lost over the decade to 2018. One third of existing veggie farmers grow their produce on less than five hectares.

It just might be that this crazy year changes that. Josh ordered his gardening equipment from James Hutchison, a supplier in Tasmania who runs a business off the back of his farm, Longley Organic Farm, half an hour south of Hobart. He seasonally supplies fifty vegetable boxes a week, which he maintains is cheaper than the equivalent basket at Coles, and he can't keep up with demand. He rejects the idea that microfarming doesn't make much difference in feeding the country.

'Food deserts are very real in city and country,' says James. 'It is about knowing farmers, and people want to know where food comes from. It's a global phenomenon. It's about reconnecting with food we are putting into [our] bodies. I don't think it's a fad, because it was definitely improving well before COVID.'

James came to food production after leaving school at fourteen and 'going through the wars'. He has been growing his business and teaching others for the past decade and says there has been a noticeable uptick in the last two years, but rapid growth has set in over the past six months. 'I think it's about self-resilience and greater food security. People being more aware of food security but also responsible for where food comes from.'

Across the country, people are trying a range of micro food-production models. James has been supplying microfarmers from outside Darwin and Alice Springs, in Adelaide, Esperance, Albany, and the Western Australian wheat belt. They are anything from small food-production systems to certified organic growers to community-supported agriculture (CSA). In CSAs, customers buy a share in the crop and stick by the farmer through thick and thin in return for food supply and connection to the farm via visits. Larger grazing and cropping farmers are also setting aside smaller plots to grow vegetables to diversify their income and test out the paddock-to-plate market.

Australia's food production – and its landscape – is changing rapidly. Three decades of economic rationalism, long supply chains, open-water markets, a dearth in federal climate change policy, the commodification of food and the concentration of a few crops means we are eroding our advantages. Disregard for food production is mining our landscapes and our human communities. COVID is forcing us to rethink all of our priorities and our daily bread might be at the heart of change.

Gabrielle Chan is a freelance journalist and the author of *Rusted Off: Why country Australia is fed up*.

Falling

JAMES BRADLEY

A week into March my mother, Denise, texted to say she was back in hospital. It wasn't a surprise: she'd been in and out of hospital for months, caught in an exhausting and often traumatic cycle of release and relapse as her cancer worsened. This time it was different, though. In the week prior she had been admitted twice already as a result of intestinal blockages caused by tumours and adhesions from previous surgeries and she was now so weak that it was obvious things could not go on as they were. The next day the doctors told her they were going to discontinue treatment. There was nothing left to be done.

Because she had been sick for so long, and had survived so many crises already, it was difficult to believe we were finally at the end.

'She's got better before, perhaps she will again,' my older daughter said when I explained the situation to her that evening.

'Not this time,' I replied.

Although she was not afraid of dying, across the eight years since her diagnosis my mother had fought hard to stay alive, continuing to work and travel and engage with friends and colleagues. But now there was no longer any way of delaying the inevitable, she just wanted to be with the people she loved. And so, as we waited for a bed to be arranged for her at the hospice, I spent many hours at the hospital, driving back and forth between Marrickville and Randwick once or twice a day.

They were unsettling days. Around the world COVID-19 was spreading fast, and listening to the radio in the car or scanning social media at home, it was impossible not to feel as if the world had stepped off a precipice. In Europe and elsewhere, infections and deaths were rising exponentially, and the scale of the economic fallout was growing worse by the moment. On the morning she was admitted to hospital the FTSE dropped by an unprecedented 7.8 per cent; simultaneously, the Saudis cut the price of oil, triggering a global economic shock. Meanwhile, here in Australia, Scott Morrison stumbled through press conferences at which he offered hopelessly contradictory advice that suggested he had little to no idea of how to proceed.

We discussed all of this in those final days in the hospital. But although she was still present intellectually, and lucid despite the morphine, it was obvious her world was contracting. And as it did, we talked less about politics and events, and more about friends and family, about values, about her life and mine. When my father was dying I thought about recording an interview with him about our family history, but it never happened, mostly because I knew he would hate the idea. One afternoon at the hospital I asked my mother whether she would let me do something similar with her. She wasn't interested. 'If you want to know about family history you need to ask your cousin Amanda,' she said. 'She knows it all much better than me.' I suspect what she really meant was none of that really mattered to her anymore.

Instead we talked about her life, the things that had shaped it. My mother grew up in Maroubra, and attended Brigidine College in Randwick. After finishing school she studied at the University of Sydney, where she was part of an unusually talented cohort. She knew Les Murray (thirty years later my grandfather was still scandalised by the fact that the one time they met Les was wearing thongs), and was friendly with Clive James and the novelist Madeleine St John, who later based the character of Lisa in *The Women in Black* on their mutual friend, Colleen. At twenty, she met my father, ten years her senior and recently returned from Oxford. When he took a position in Adelaide she followed him, and they married in secret; my grandfather was so angry when he found out he tried to have the marriage annulled.

Later, my grandparents would move to Adelaide as well, but her relationship with them was always complex. Shortly after my mother's younger brother Richard was born, my grandmother suffered an embolism in her leg, a medical catastrophe from which she never recovered. For the rest of her life she walked with a stick, and was in constant agony from the swelling in her lower leg and foot.

In the aftermath of the embolism my grandmother began to drink heavily, and continued to do so right through my mother's childhood and adolescence. Finally, around the time

I was born, she was admitted to Chelmsford Hospital, where she underwent several weeks of deep sleep therapy. These days Chelmsford's name is synonymous with death and disaster, but my grandmother's time there was an unexpected success, and she emerged sober and remained that way until my grandfather died twenty-five years later.

Growing up, I was always acutely aware of my grandmother's physical vulnerability; only later did I recognise the anger and frustration that had shaped her life. But while I never doubted she loved me, my mother's relationship with her was much more difficult, and far less resolved. Although we had talked about it before, she said in those last weeks that my grandmother's alcoholism meant she arrived in adulthood without any real model for how to be a parent. And so, when I was born (followed, relatively quickly, by my three brothers) she found she had to learn to be a mother, a process she had found far harder than she had expected, but ultimately transformative.

Being a parent also taught her to be more disciplined. 'Having kids forced me to grow up,' she often said. She started her working life as a high school teacher (on two-thirds of the pay of her male colleagues because she was married), before taking up a position as a teacher-librarian because she could fit it around minding me and my brothers. She then went on to teach librarian studies at teachers' college. By the time I was in high school she was dean of her faculty; by the time I was in university she was a vice-chancellor and a leading figure in Australian higher education.

These achievements didn't happen by accident. She was formidably intelligent, and possessed a remarkable ability to synthesise vast amounts of information and think strategically, as well as a passionate belief in the importance of equity and education. She also worked harder than anybody I have ever met, and even in retirement was working more than most people do when full-time. Less than a month before she died and already on high doses of morphine, she attended an all-day board meeting from her hospital bed. When I asked her how it had gone she told me it was fine, except that they'd made her have a brain scan in the middle of it.

I suspect that for many of us, those first weeks of the pandemic already seem like a strange and vivid dream. Not just because of the confusion and the panic, but because of the eerie, almost uncanny normality of it all. One afternoon on the way home from the hospital I bought a coffee and sat outside a café near my home. The global death toll had just passed 6000, and South Africa had announced it was closing its borders. Yet sitting there, in the sun, so tired I could barely think, there was no sign of any of that. Instead it felt like there was a tsunami heading towards us, a disaster we could see approaching and knew we could not outrun, but which had yet to strike. And, as the world began to buckle and shift, it had become harder to ignore the fault lines running through our society. Many of these had been apparent for some time: the corrosiveness of a culture that privileges self-interest and material wealth over community and sustain-ability, the growing alienation and anxiety caused by economic precarity and technology, the steady erosion of the civic institu-tions that sustain our lives. But the looming catastrophe of the pandemic had thrown them into even starker relief. One night after I left the hospital I went to the supermarket to buy some milk. It was raining outside, and the supermarket was crowded, its shelves stripped bare of pasta, tinned food and toilet paper. I remember standing by the checkouts and watching the people around me, the way they kept their eyes down, avoiding each other's gaze, suddenly aware of how fragile this all was, how easily it could go down.

Eventually, a place for my mother became available at the hospice. Despite its finality, we thought this was a good thing: although she had been comfortable and felt safe in the hospital, she needed specialised palliative care. But when we arrived we discovered the hospice had just instituted new rules restricting each patient to one visitor a day, with each visit limited to a maximum of fifteen minutes.

That first morning we argued our way past reception, but that evening the staff told us that while they weren't going to enforce the time limit, we couldn't have more than one person in the room at once. When we tried to abide by that rule by doing shifts in her room, the nurses made it clear they weren't happy about us sitting in the hospice café while we were waiting, and later that day one of us overheard a doctor telling nurses it was time to start enforcing the fifteen-minute rule. Since my mother was already confused and distressed, this was not a situation any of us could countenance.

The rolling uncertainty about our ability to care for her as we needed to took its toll on all of us, but it was hardest for my stepfather, Bruce. Already exhausted by months of grief and worry, he was extremely concerned that the hospice might ban us altogether, and terrified she would be alone when she needed us most.

There were other problems as well. My brother David had flown up from Adelaide the week before, but with each new day the travel situation became more and more uncertain, as countries around the world banned international arrivals and airlines declared bankruptcy. On my mother's third day in the hospice, Tasmania announced it was closing its borders and the other states and territories announced they were considering similar restrictions; that same afternoon the palliative care doctor told us that despite the fact she was only intermittently conscious, she might survive another week or even two. Faced with the prospect of being unable to get home, David elected to stay, but it wasn't an easy decision. The next day I had a similar conversation with my uncle Richard, who lives outside Gympie in Queensland. He'd spoken to my mother on the phone the week before, and wanted to see her one last time, but was now worried about the risk of infection if he got on a plane. I told him that even if he came she probably wouldn't know he was there. He didn't get to speak to her again.

My mother died the next morning. Only eleven days had passed since she had been admitted that last time, but it already seemed like forever, or no time at all. After we had tidied her room I went home to see my partner, Mardi, but once I was there I didn't know what to do with myself. I didn't want to be still, but it was too hot to go walking, and the thought of doing anything normal seemed not just surreal, but somehow wrong. I am familiar with the way loss dislocates us: the day after my father died I went to my friend Malcolm's fiftieth birthday party. I realise now I was still in shock, deranged by grief, that although I talked and laughed, I was somehow outside myself, not really there.

This was not quite like that. Instead, it seemed as if she wasn't gone at all, as if the world had simply stopped, and she was still out there somewhere. It's possible this was partly because we had all been grieving for so long already: as a friend said a few weeks later, often the pain of losing somebody begins long before they actually die. Perhaps it was also because there had been so many crises along the way, so many times we had prepared ourselves for her to be gone only to have her recover against the odds. After the worst thing has almost happened so often, sometimes it no longer feels like the worst thing.

But it was more than that. As I tried to make sense of her sudden absence, the world was unravelling. On the day after she died, still numb with grief, I watched images of army trucks transporting the dead to mass graves in Italy and Spain, while here in Australia governments announced school closures and the police ordered swimmers off Bondi Beach. Every hour, every minute, brought some new and usually terrifying development. In the face of all that, my private grief seemed if not irrelevant then somehow incommensurable, impossible to parse or process.

Perhaps it would have helped if we'd been able to hold some kind of ceremony for her. But with funerals effectively banned and my brother David and stepsister Laura back in South Australia with the borders closed, that was not possible. Even seeing my stepfather, now alone in their apartment, was incredibly fraught: although we had decided quite early on that his need for company and comfort outweighed the risk of infection, every visit to him was charged with the knowledge we were breaching the rules.

A psychologist friend of mine talks about the idea of frozen grief, a phenomenon that occurs when people are denied the normal communal rituals associated with grieving, meaning that their feelings cannot be expressed or processed. And indeed, as the weeks passed, I found myself adrift in a hinterland of loss, a strange no man's land I could not seem to escape. In the evenings, after we had finished trying to work and school our kids, Mardi and I would walk by the Cooks River. By the time we got there it was always dark, the streets deserted; often it felt like we were the only people left in the world.

Grief is always isolating. It cuts us off from the world, confines us in ourselves. Yet, as I watched the people I know on social media and elsewhere trying to express the confusion they felt at being pitched into a world where they were suddenly vulnerable and alone, it was hard not to wonder whether this wasn't also a kind of frozen grief. Like the climate crisis, the pandemic has altered time, making it elastic, permeable, destabilising our futures and erasing the boundaries between past and present. Our senses of loss and isolation, of uncertainty and fear, are themselves forms of grief, often inchoate and inexpressible, but real, and profound.

I want to write what my mother meant to me, about the way her honesty and her belief that the world can be a better place has shaped my life. I want to write about my love for her. But I don't know how, not yet. Because like all of us I feel undone, unmade, as if time has been suspended, and the world I know is gone. As if I am falling, and have not yet hit the ground.

James Bradley is an author and a critic. His books include the novels *Wrack*, *The Resurrectionist*, *Clade* and *Ghost Species*.

Call and response

CHRISTOS TSIOLKAS

For my parents

As soon as my partner and I completed two weeks' quarantine after returning from Europe in late March, I went around to visit my mother. We had talked on the phone every day throughout the lockdown, and twice during our quarantine she had defied the warnings against elderly citizens leaving their homes to drive to our place and deliver food parcels. There are images of the pandemic that will forever be part of our cultural memory: the long lines at Centrelink after thousands of Australians lost their jobs; the furious scrambling for goods in the supermarkets; and the desperate faces at the windows of those poor souls in Melbourne's public estate towers who were sent into cruel and immediate detention when the city lurched haphazardly into a second lockdown. But every single one of us will have a unique memory that will define the staggering dislocation that we all experienced in 2020. For me, it is my mother standing outside our house. I am at the window, waving at her and mouthing 'thank you' for the food. And she is weeping. She so wants to embrace and welcome home her child. COVID-19 still makes that impossible.

Negotiations and calculated risks: they become second nature over the period of the pandemic. My mother is 82 years old and the health advice is that she should stay at home. My brother and I argued with her on the phone, explaining that the protocols are for her own good, that her demographic is most at risk. She has read and listened to the advice in the Greek-language media but she decided to continue driving and doing her own shopping while maintaining the rigours of social distancing and sanitation. After my third attempt to convince her to stay home, she snapped at me.

'I refuse to be a prisoner. Let the Lord take me if He wants. I need to breathe!'

She pleaded with me to visit when I was free from quarantine. I drove over to her house and we sat at opposite ends of her kitchen table. At one point, as we were deep in discussion, she leaned forward and I gauged that it was her instinct to touch my hand. I sprang back in my seat, knocking the chair against the wall.

'My Lord,' she exclaimed, hugging herself tight as if to reign in the humanity of her impulses, 'I've lived through hunger and Occupation and civil war but even in the worst of times we were allowed to touch, we were allowed to hug each other.'

And then, quickly, she crossed herself.

'We're lucky,' she said. 'We're not ill and we're not homeless. Let's be thankful for that.'

★

That afternoon she asked me about being in the UK as the whole world stumbled punch-drunk into lockdown. I tried to describe the astonishing disorientation we experienced while navigating the crowds at Heathrow, Dubai and Melbourne airports. For the first time in my life I was disturbed by the cavalier assumption so many of us had that there was nothing extraordinary in assuming the right to air travel. We were women and men. We were young and old. We were black and brown and Asian and white. And we were from across the globe. Why were we all taking this miracle of flight for granted?

I made her laugh when I described the young woman sitting next to me at the back of the economy section. That woman was cheerful and loquacious, but from time to time was wracked by a terrible cough, and she would huddle herself into a ball to make sure she coughed into her armpit. At one point, seeing her distress and her guilt, I tapped her on the shoulder and pointed around the crammed cabin: people in line waiting for the toilets, a mother nursing her crying child, the harried stewards trying to navigate the meal trolleys in the narrow aisles that made nonsense of the pilot's take-off pleas for maintaining social distancing.

'It's okay, if there's COVID on this flight, we're all going to get it.'

My mother and I both laughed.

'Who knows when I will fly again?'

My mother shrugged at my question, and then, turning, she pointed to the myriad postcards Blu-Tacked on the kitchen cupboard doors that I had sent her over the years. India and Mexico, Turkey and Jordan and Spain, New York and Tokyo and Paris.

'You've travelled enough.'

It wasn't an admonishment; it was a simple statement of fact.

My father, who migrated on a ship to Australia in 1955, returned twice to his homeland in Greece. The first time was in the early 1970s to farewell his father who was dying, and the second time was in the early 1990s to bury his mother. It is only very recently that I realised the immense financial cost of that first trip back home. My parents were saving up for a new house but when he received the call that his father was ill, they sat down together and made the decision that it was his duty to return home. Everything else would have to wait. At that point, air travel was a luxury and even though my father organised the cheapest flight possible, one that took him from Melbourne to Manila and from Manila to Karachi and then to Athens, my mother's recollection was that it still ended up costing $2000. In 1972, that money would have been a sizeable deposit on a house. It was not conceivable that we could go as a family. My parents were labourers in inner-city factories. The costs would have been prohibitive.

In the near fifty years that my parents were married, they never flew on a holiday together. I looked at the postcards in my mother's kitchen, those pretty vistas of Europe and Asia and the Americas, and I felt shame.

For most Australians, two events will clearly define 2020, and they are events imbued with an atavistic, Biblical solemnity: fire and plague. The new year began with the catastrophic fires that destroyed lives, communities, towns and forests all down the eastern seaboard of Australia. Yet, amid the distress and mourning, the evaluation and the taking stock, I found myself hopeful. I travelled up to the NSW South Coast and was humbled by seeing a large group of firefighters emerge from a smouldering forest, their faces scorched by the detritus from the fires. They were heroic and they were noble, and I didn't give a damn whether they individually voted Labor or Coalition, Shooters Party or Greens. And there were so many people wanting to contribute money, assistance and resources to those most affected by the fires. These expressions of unity and goodwill, of compassion and sorrow, seemed to cut across the ugliness of partisanship and hectoring outrage that has defined our politics for so long. It even seemed possible that, after a decade of political bickering and inertia, we would finally as a nation take definitive steps to address climate change.

When I reflect on that time in January – only a few months ago, but the subsequent arrival of the COVID-19 pandemic makes it seem like another age – I realise that my optimism was naive. The idiocies of right-wing culture wars and the equal absurdity of left-wing identity politics soon re-emerged to clog the sewerage that is social media. Nevertheless, there was something undeniably sobering for all of us in acknowledging the terror caused by the fires. We had to contemplate tragedy. There are many Australians who have known tragedy already: generations of refugees and migrants who fled wars and dictatorships to create new lives in this country; and the First Australians, engaged in the too-long struggle to unshackle themselves from the bondage of dispossession. There is individual tragedy, too, of course: the loss of a child or the loss of a partner. But it has been a long time since we have faced tragedy *collectively*.

The fire and the plague are a call and response, a motif I borrow from music, but which seems to me appropriate for this time. The bushfires of last summer and the pandemic of this present moment are two distinct phenomena. Yet they have occurred in a specific moment in time when accelerating globalisation has undermined the centuries-old legitimacy of the nation-state. These historic forces have bound together the fires and the pandemic. But the fire and the plague have disrupted the exultant progress of globalisation.

Most of us understand that combating climate change requires an international response. However, the severity of the fires also made us understand that a continent such as Australia will face unique challenges from a warming climate. And if you were stuck in an airport in mid-March of this year, as the immensity of the COVID-19 crisis dawned on the whole of the world, as you tried desperately to plead for a ticket on a flight back home, you would have realised that no other element of your identity was as important as the stamp on your passport. I was in Europe but my Greek heritage mattered naught. Borders were slamming shut at breathless speed. Fuck off back to where you came from. The messaging was as succinct and as clear as that.

The shame I experienced looking at the postcards I sent my parents was ignited on that wretched flight home from Dubai to Melbourne. Looking around at that overcrowded cabin, I recognised that, for all my smug self-belief that I was on the winning side of history when it came to climate change, the many flights that I have taken over the last few decades – from Melbourne to the world, from Melbourne to Sydney, to Brisbane, to Alice Springs, to Perth – have contributed so much more to carbon emissions than if I had been living in an air-conditioned

McMansion at the outer suburban edge of my city. Call and response. One voice declares, 'I am righteous and cosmopolitan and a compassionate citizen of the world.' The other voice rises and asks, 'Who the hell are you to throw the first stone?'

There is a cohort of my friends – people I love and adore, with whom I have partied and worked and cried – who sometimes lament my allegiance to family. Strongly feminist and proudly progressive, they are suspicious that my respect for family tilts me towards conservatism. They are probably right. The last few months I have been describing myself as a conservative socialist, wary of the moral shapeshifting required from contemporary progressivism. The importance of family for me clearly has to do with the symbolic and material consequences of my parents' migrations. In leaving their country at a time of war and poverty, in forging a new life in a new country in which they did not speak the language and which was on the other side of the earth, my mother and my father endured hardships that are inconceivable to a soft generation like my own. Their call is sacrifice. My response *has* to be gratitude.

But it isn't only about migration. It also has to do with class. My father's knees were destroyed by the long years on assembly lines, as were my mother's wrists. Soon after coming out of quarantine I caught up with an elder distant relative. She has spent her life here in Australia, working in factories and in retail and as a cleaner, and though she admits that she never reads – the literary world is another planet for her – she has always expressed pride in my being a writer. Seeing her again, we spoke of the fires and we spoke of the pandemic. She's never asked anything of me. But she did this time. We spoke of what the world was like before COVID-19 and she sighed and said, 'It's hard. There's no security in the work my children do, everything is too expensive, and there's no faith in politics.' We both lamented the pace of globalisation. And then, shrugging, I said, only rhetorically, 'What can any of us do?'

That's when she said, 'There's nothing I can do, Christo. But you're a writer and you have a voice. Tell them how hard global-isation is for us. Tell them we need industries and jobs in this country. Tell them we need a future and tell them we need dignity.'

Her call.

My response?

Before Victoria went into a second lockdown, I dared to catch a tram again. I had my hand sanitiser in my jacket pocket, I made sure to wear gloves. Behind me were three young women, VCE students from the inner-city high school down the road. They started discussing the Black Lives Matter rally they had all attended the week before. My ears pricked up. For one brief moment, I had contemplated going to the protest. If I'm truly honest, not for the murder of George Floyd. That murder was sickening and wretched and unjust; but it happened in the USA and if this pandemic has taught me anything it is that, contrary to a belief that I myself have espoused in the past, we are not the USA. *Thankfully, we are not the USA.* But the unconscionable rate of incarceration of Indigenous people and the continuing blight of Aboriginal deaths in custody gave the rallies in Australia a clear moral purpose. However, I knew that if I did attend, I would immediately choose to return to a fort-night's quarantine. Given my responsibilities to my mother, and given that my partner's mother is also elderly, I decided against going. The young students were passionate about anti-racism, and they dropped the word 'privilege' a few times. But very soon their conversation turned to how COVID-19 had ruined their plans for schoolies week. They had been planning to fly up to Byron Bay but that now seemed impossible. 'It's so unfair,' one of them sighed.

Did I speak up? Of course I didn't. I remember being young. But I thought of the many gifts my parents had given me: the gifts of patience, of compassion, of sacrifice. I realise now that one of the greatest gifts I received from them is the gift of gratefulness. My mother crosses herself and declares, 'We're not ill and we're not homeless. Let's be thankful for that.' It was in the interplay between gratitude and the responsibilities that ensue from that understanding that my ethics and my politics were formed. That gift is my bedrock.

I recall the faces of the weary, dirty firefighters emerging from the fire-scorched bush at the start of this calamitous year. I remember the chaos at the airports in March and my dawning realisation that our reckless indulgence in travel might be over. I think of that pretty row of postcards.

So, my response to this year of plague and of fire is to offer you the gifts my parents gave me: now is the time for patience, for compassion, for gratitude. Now is the time for sacrifice.

Christos Tsiolkas is a celebrated author, playwright, essayist and screen-writer. In 2020 he won the Prize for Fiction at the Victorian Premier's Literary Awards for his most recent novel *Damascus*.

Friday the thirteenth

MELANIE CHENG

As a teenager, I loved feeling scared. Horror films were my go-to. *The Exorcist, Poltergeist, The Omen*, even the slightly sillier ones like *Friday the 13th*. In my youth, I mistook that manufactured titillation for real fear, but now I know better. Now I know true fear is not exhilarating. True fear cannot be easily soothed by a quick cuddle from Mum. True fear is intense, exhausting, merciless. True fear is an invisible pathogen that threatens to strip you of everything you love.

'Would you prefer to get your results on a less ominous day?' I joked as I booked patients in for their appointments on Friday the thirteenth of March. Some laughed, and others hesitated, but most seemed to eventually suppress any niggling superstition.

I can't say it began like any other day. It was March 2020, after all. Things hadn't been normal at the clinic for several weeks. I, like other general practitioners across Australia, had been revisiting skills I hadn't used since my hospital residency: isolating patients, taking nasopharyngeal swabs, donning and doffing personal protective equipment. Even the terms felt awkward, antiquated, ridiculous. More like tongue twisters than descriptions for lifesaving precautions.

Just three days prior, on 10 March, Australia had achieved the portentous milestone of its hundredth case of COVID-19. Around the same time, courtesy of the medical grapevine, descriptions of the situation in northern Italy were starting to sneak in. Doctors spoke in hushed tones of the impossible decisions colleagues on the other side of the world were having to make. Decisions like whether to give the last ventilator to a forty-year-old father of one or a forty-year-old father of two. Entire lives were being reduced to cold hard demographics: the number of dependants, the number of comorbidities, the number of years a person had left to live. According to the experts, Australia was approximately two weeks behind Italy's trajectory. For those of us on the frontline, the situation felt less like real life and more like a scene from Neville Shute's *On the Beach* – a group of survivors nervously awaiting the arrival of a deadly cloud.

We kept busy. In the absence of a proven treatment or vaccine, we went back to basics: an effective triage system, isolation of high-risk patients, mandatory masks, and a healthy dose of vigilance. Unfortunately, I overdosed on vigilance. I woke with a start at 4 o'clock every morning, and before my eyes had even adjusted to the dark, I hunted for my phone on the bedside table to check the latest coronavirus numbers. I listened to every podcast and tuned in to every webinar in an attempt to arm myself with knowledge. But I wasn't a junior resident

anymore, I was a GP who'd been around long enough to understand that knowledge doesn't always translate to control. The sad
truth was, medicine, with all its fancy technology and billions of
dollars of research, could offer nothing more to protect me than
a plastic apron and a duck-billed mask.

I dreaded going to the clinic, which was unusual for me.
Not since my internship had work filled me with such terror.
Less than a week before, a Melbourne GP had been named and
shamed in the media for attending work with what he thought
was just a cold, only to subsequently test positive for COVID-19.
Whenever I stopped long enough to think about whether I, too,
had viral symptoms, I felt an ache across my back and shoulders,
a dryness at the back of my throat. One day, at school pick-up,
after a day of testing patients for coronavirus, I startled a friendly
mother who had leaned in for a chat by holding my arm out
long and straight, in the manner of traffic cop, and warning
her to stay away. It was still early days – apart from those of us
working in health care, most people were still going about their
usual business. Schools, cinemas, theatres, sporting venues and
universities were all still open. Even I, who arguably should
have known better, had plans for a chock-a-block weekend, full
of soccer training, birthday parties and catch-ups with family.

On Friday the thirteenth of March, it seemed as if everyone
was sick. Many of the patients I saw that morning had upper
respiratory tract symptoms. Unfortunately, none of them met
the criteria for COVID-19 testing. All I could do, in addition
to the usual fluids-rest-paracetamol advice, was tell them to
go home and stay away from others until they'd completely
recovered.

As the morning wore on, the dryness at the back of my mouth progressed to a sore throat. During a tea break, I received word that the fortieth birthday party scheduled for Saturday night had been cancelled. The hosts didn't want to be responsible for an outbreak among loved ones. I felt relieved, and then thankful to have such cautious friends.

At lunchtime, after I'd finished writing my notes, I checked the latest news on my phone. Nine new cases of coronavirus had been detected in Victoria, including, significantly, the state's first case of community transmission. As I scrolled down to the list of exposure sites at the end of the article, I felt a sudden churning in my belly. With disbelief I read the name of my favourite café, where my kids and I went for coffee and sweet treats almost every day.

This thing, this germ, which had only a month ago seemed so far away, was now circulating in my community. This fact triggered a cascade of unbidden and unhelpful thoughts. What if it wasn't a stranger at all, but me, or a member of my family, who had brought the infection into our neighbourhood? My children had both had fevers a few weeks before. They'd recovered, but now I had a sore throat. Was it possible I had been unknowingly incubating the deadly coronavirus and passing it around to my friends, my parents, my in-laws? Was I the next doctor to be vilified in the press?

After frantic phone calls to my husband, my manager and the Department of Health and Human Services, I elected to get tested. With a tremulous hand, I inserted a swab so far back into my nose it could have been a party trick if anyone had been around to witness it. As it was, I was alone in my consulting room. It was quiet and anticlimactic. But the significance of the moment was not completely lost on me – I had the sense that I was about to launch an unstoppable chain reaction. There was no going back. I sealed the swab in a small ziplocked bag, dumped it in the bin for the courier and went home to wait for my results.

That afternoon, the prime minister announced a ban on public gatherings of more than 500 people. He encouraged us not to panic. Keep calm and carry on, he said. But, as someone who has treated patients for anxiety, I knew that simply telling people not to panic wouldn't be enough. It certainly didn't do much to relieve my anxiety. The forced isolation, however, did offer some respite – at least if I was at home, I couldn't infect anyone else.

My next hurdle was telling the kids. Coronavirus may have surpassed Voldemort and Darth Vader as the supreme baddie in their schoolyard games, but they had not yet come to expect the disease to impact their lives in any meaningful way. My daughter took it the hardest. 'It's true what they say about Friday the thirteenth!' she yelled when I told her she wouldn't be able to go to her friend's birthday party that weekend. 'It's the worst day ever!'

My test results came back on the Monday. Coronavirus was not detected. It was a relief. Unfortunately, my relief quickly turned to despair. In the week that followed, the number of confirmed cases of COVID-19 in Australia doubled every three days. Our curve started tracking dangerously close to Italy's and the USA's. In spite of the PM's entreaties not to, people panicked. Supermarket shelves were stripped of everything from flour to cans of tuna. As neighbourhoods prepared for lockdown, sales of televisions, iPads and laptops soared. At work, masks, hand sanitiser and even thermometers started disappearing from our trolleys. More patients than ever before presented to the clinic for testing. My email inbox was bombarded with updates from medical organisations – all the usual ones and some I'd never heard of before. I felt as if I were being buried alive by an avalanche of information.

Perhaps the most confronting experience of all was performing nasopharyngeal swabs on suspected cases. As I stood in the isolation room, taking as brief a history as I could in order to minimise my exposure time, my sympathies lay heavily with the patients. How must they feel, being tested for a potentially fatal disease by a human being wrapped in plastic? I tried to comfort them by being as expressive as I could through my masked mouth and goggled eyes but I knew it was hopeless. I took the swab and got out.

At the end of the week, my husband called me as I was driving to work. He'd just received word that three anaesthetists in the UK had been admitted to the ICU with coronavirus. Only days before, as a consultant anaesthetist himself, he had 'opted in' to intubate COVID-19-positive patients. Neither of us mentioned the kids during that conversation but we didn't have to. They were there, in every pause, in every weighty silence. When I arrived at the clinic and a colleague asked me how I was, I dissolved into a mess of hot, ugly tears.

In the months of lockdown that followed, I, like everyone else, watched the daily coronavirus numbers. My anxiety rose and fell, like flotsam, on the wave of that ominous graph. Finally, sometime in late April, I reached a point where I could sleep, laugh, read again. But the reprieve was short-lived. Now, as the Victorian numbers swell once more, so too does my fear. Only this time the source of my terror is less nebulous than before. Now the bogeyman has a vague form and features – he takes shape in the empty streets, the shuttered shops, the panicked voices of patients.

When I was a kid watching horror movies all those years ago, I confused the exhilaration I felt with true fear. But really, it was just a watered-down version of it – as close an approximation to true fear as fruit punch is to absinthe. I always knew the ghosts in the films weren't real and couldn't hurt me, just as I knew my parents were in the next room, at hand to reassure and protect me.

The pandemic has shown me that true fear occurs when we feel that someone or something has taken control of our fate. It occurs when we contemplate our nonexistence; when we acknowledge our fragility, our insignificance. We've all had experiences like this. A friend goes to work and doesn't return home. A family member is told a cluster of cells somewhere in their body has gone rogue. In these watershed moments, our little worlds are turned upside down. But usually, the wider world carries on. And while we may find this fact incomprehensible, even offensive, in times of personal tragedy, there is also some comfort – and hope for our own recovery – in seeing everyone else keep going.

What is so exceptional about a global pandemic is that the wider world does not carry on. When we stop, the wider world stops too. And in the chance moments that we meet – at a safe social distance, wearing masks – and catch each other's gaze, we see nothing but our own fear reflected in a stranger's eyes. It is then that the illusion we have all worked so hard to maintain for so long – that we are supreme and all-powerful and important – finally falters. The jig is up. We were never in control. And there is nothing quite so terrifying and transformative as that.

Melanie Cheng is a writer and general practitioner. She was born in Adelaide, grew up in Hong Kong and now lives in Melbourne. Her debut collection of short stories, *Australia Day*, won the Victorian Premier's Literary Award for an Unpublished Manuscript in 2016 and for Fiction in 2018. Her first novel, *Room for a Stranger*, was published in 2019.

The music of the virus

BRENDA WALKER

One fine untroubled morning in 2019 I was out walking in Potts Point, on my way to see my eldest brother. He lived in a room here when I was twenty-one and he was twenty-six. In those days, Potts Point was unconventional and impoverished, home to people who minded their own business, which was largely conducted at night.

His room was narrow, with a bed, and a wardrobe housing a few shirts on wire hangers. A window opened on to a wall. There was a bathroom on the same floor. I could stay there when he was away; I could borrow a shirt. When he wasn't away I stayed in a friend's apartment on New South Head Road and walked to Potts Point to visit him. At the time, I was writing a thesis on the fiction of Samuel Beckett. As I wrote I grew more and more uneasy about the loss of this thesis, and I began to carry my work with me in a small suitcase for safekeeping. With my suitcase and my plain man's shirt I wasn't of much interest to the people on the street. I kept writing. The suitcase became heavier and heavier, for it now contained books and all my drafts. I carried it to my brother's concerts. We began to share this burden, as we walked about the city. Once he stopped and put it down, flexing his fingers. 'You do realise I make my living with my hands,' he said, before he picked it up again.

Sometimes I met him at the recording studio, a room with a strange artificial quietness. The master tape we listened to was more distinct, clearer than the mass-produced sound it created. I have often thought of a family in this way: a space where the wider world is lightly muted. What happens there can have a particular irreplicable clarity.

Potts Point is prosperous now. In 2019 I walked past tubs of flannel flowers and crimson waratahs on my way to my morning reunion with my eldest brother. But that afternoon I had flown over a desiccated landscape to reach the town where we once lived. I had read about the drought, but I hadn't expected to see this: bare earth, trees white as bone, sheep, their coats weighted with dust, gathering at dams to drink from milky, unhealthy water. The walls of the dams were ridged at points where the water level had held for a time. Later, much of the land I had seen was overwhelmed by fire. Millions of hectares of precious forests, some ancient and irreplaceable, were destroyed, before immense human effort and unforeseen rain extinguished the fires.

The country was still raw when we heard about the virus. By that time, I was back in Western Australia. Reports of illness and suffering in China were concerning, but remote. Then the virus appeared in other places: France and Italy, London and New York. It was spreading in Australia. We were told to stay at home. With images on our screens of prone bodies in the rigging of intensive care units, of unemployment queues, of the pragmatic interiors of open graves, it was hard to remember that there had been such a thing as an untroubled morning in a fine city.

Most years, in the early hours of 25 April, I wake to the sound of a low murmuring from the streets below. Our apartment building acts as a sea wall, and on this particular morning the sea that flows beside it is made of people walking, measured and purposeful, up the hill to the war memorial that sits at the crest of Kings Park. They take up position early, in darkness.

2020 was different. The Anzac Day commemoration was cancelled because of the threat of the virus; there were no pre-dawn voices. Later that morning, when my husband and I walked through the park where the war memorial stands we saw large signs telling us to stay at home. The park has been an Indigenous dreaming place for tens of thousands of years. The war memorial lies above the meeting of wide rivers: a place of sky and water. There is a memorial to the Boer War, but no memorial to the Frontier Wars. In 2020, despite being told to stay away, people had left behind twists of rosemary and garden flowers. I thought of my husband's mother, a thin child in dark photographs, born in Novgorod during the Russian Civil War. We paused under a familiar tree. The coffee kiosk was open.

The signs said stay at home. Our home is a shelf of concrete walled in glass. It sits above the tree line, below the steady flight path of herons. Pairs of magpies and parrots rest briefly on streetlights. Below them, another understorey of swift insect eaters rushes closer to the dangerous human activity of rushing bodies and machines. I notice a kind of courtesy among the birds, a sharing of space, but there are also disputes. Some birds land on our balcony and look in on us; others menace them. Small birds follow larger egg and nestling eaters, attacking their flight feathers. Ravens, in particular, are hated and pursued. The air below the herons is full of purpose, of danger and avoidance. The virus brings us closer to this: we are shut in, attentive. Stay at home, with the wild birds and the TV screen showing hospitals and queues and graves, with reflexive cooking that no one has the appetite to eat, with whatever music might carry the gravity of what is happening and still rise through the air.

In the opening scene of a BBC report from a hardworking hospital in East London a man in blue scrubs played a black piano. This crowd-funded piano sits in the hospital foyer. The shot was beautifully composed: we heard such calm proficiency as notes rippled forth from conspicuously medical hands. The physicians, the nurses, the cleaners might all be gifted with hands as dexterous, if not as musical, as these. Then there were interviews about the virus. A consultant surgeon said of the patients and staff, 'There is kindness everywhere.' In my mind, the kindness and the piano are linked. Pianos are paradoxes: instruments of solitary absorption and memorised sequences of sound, but also mighty sources of communal consolation. In the aftermath of the killing of George Floyd, a piano appeared at the scene of his death and people began to play.

When Beckett was a student in Dublin, the man who shared his rooms overheard him improvising on a rented piano: sad chords, solitary and nocturnal.[1] Beckett played the piano for most of his life, but he didn't own one until he was sixty-one.

It was characteristically modest, a German Schimmel, installed in the equally inconspicuous country retreat where he wrote, drank whiskey and isolated himself by choice. He was a physical player; described as 'pounding' the keyboard of a piano at the École Normale in Paris.[2] His eyesight was so bad that he had to contort himself for sight-reading; 'my nose so close to the score that the keyboard feels behind my back.'[3] We see him leaning in to the piano, tipping forwards like an old acrobat, joking about it later.

When an early manuscript of his novel *Murphy* was sold by Sotheby's in 2013, some pages were displayed online, and among his incidental doodles – a portrait of James Joyce and another of Charlie Chaplin – he repeatedly drew the treble clef: a twirl on the page, a spiral of impulsive gorgeousness. This is also the thing the piano can do: tip us into a place of delight, take us somewhere where we leave the weight of our anatomy behind. Moneyed refugees fleeing the Russian Civil War valued their pianos so much that they strapped them on top of the trains that carried them through Manchuria and out of their country.[4] But even when pianos must be left behind, the notes still travel, contained in the mind. My husband's mother, twice a refugee, having put aside her musicianship and most of her languages after arriving in Australia, surprised her family with a single flawless piano performance in her seventies.

Lately, I have been dreaming of a piano of my own, an instrument for sadness and relief. I have been dreaming of a piano that I cannot see. The keys are visible, and so are my hands, but the rest of the instrument is indistinct. In dreams before my son was born I was given a swaddled baby that I could hold, provided

I didn't uncover his face. Perhaps this piano dream has drifted across from my old dream of my unborn child.

On our screens we see hospitals, queues for food and work, the interiors of homes in New York and Italy, glimpsed as cameras film the removal of stricken bodies. We see the bedding, the shrines and decorations of strangers: their domestic consolations. We see rows of graves and look away. My friend in the apartment above texts me, asking if I can go to the far corner of our balcony and look up at her and wave. She's standing at her window, framed like a portrait. She hasn't been out, nor seen another living person, for a week. We laugh and shake our heads at this peculiar meeting. I see my grown son in a carpark, we catch our breath and hug, our masks secure.

People are finding ways to pass the time. Virtual museum tours, concerts from musicians' living rooms, podcasts, Netflix. Everyone, it seems, is ordering food deliveries. I discover that removalists are still at work. Soon I will find a small piano on the internet and men will manoeuvre it into our bleach-smelling lift and wheel it through our door and into position. Curious birds will watch them, briefly, from the railing of our balcony. When I begin to play, so much memory will be released. My own first chords, I know, will be sad.

In the meantime, I am listening to the sonic version of the virus, created by Markus Buehler at MIT. It sounds elegant, pizzicato, like the koto music I sometimes listen to when I'm writing or cooking: background music that settles the mind. Perhaps American scientists are habitual listeners to koto music as well, standing at their desks or kitchen benches, hands busy with a keyboard or a knife, minds adrift. Perhaps when Buehler was considering the patterns in the topography of the virus, when he was assigning instruments to elements of these patterns, he chose flute and strings but gave primacy to the koto, that calm companion of the mind. This sound brings the virus into my home in an unconvincingly neutral, abstract form. I listen, I watch the birds; I cook for our fading appetites. I think about the past.

The friend whose apartment I used to stay in on New South Head Road when my brother's room was occupied had twelve Siamese cats. These cats kept vigil over me in the night; if I woke, I would see them perched on my Beckett suitcase and the armrests of the couch where I slept. They were pale and clean-limbed, staring at me with gentle puzzlement. Sometimes I spent the entire day in the musty old Art Deco apartment, writing, my suitcase open and the contents strewn around me. That suitcase, which must have fallen apart in time, is within me in ghostly form, filled with all manner of notes and memories now, immune to the pathologies of the world outside.

We resumed our normal lives. Then my husband was diagnosed with cancer. Each day I walked between the hospital car park and the ward where he lay with other, much older men. The private rooms had been left vacant for COVID patients. Our marriage shrank to a narrow curtained space, a metal bed with railings. We heard conversations from neighbouring cubicles and the constant sigh of the machine that supplied rhythmic compressions to his immobile legs to prevent thrombosis. Our own voices could be overheard as well, so we barely spoke. Kisses are silent, touch is silent, hands fit together in silence, a woman watching a man sleep is also very quiet.

I cried in the hospital corridors, which were long, occupied by patients struggling with post-operative attempts to walk or orderlies wheeling their long trolleys between wards and operating theatres. Nobody noticed me. Mindful of the virus, people preferred not to share a lift. The pavement outside the hospital was marked with directions to the COVID clinic. Alone at night in our apartment, something happened to my hands. There was no loss of sensation, but if I brought them together they didn't recognise each other, it was as if I was touching another person.

Someone I barely knew organised a concert for me. The pianist was a friend of his, a man I had never met. It was held in a performance space with a polished wooden floor, dark drapes and an overhead light shining on a vast black piano that produced a sound that moved through my body and briefly remained there, vibrating, as if my very bones were simply adjuncts of the ear. Apart from the pianist, I was alone.

I hoped that when I found my own piano my hands would remember they belonged to one another.

Because of this virus, we all have a sense of what it means to live in disturbing times, to live under threat. Threat can take so many forms: one of the more obvious is cancer. We should also not forget the people who have experienced threat and disturbance all their lives. When I was twenty-one I often made my way up to Potts Point where old men, relics of orphanages and prisons and hostels, rested against walls or sat on ledges in the sun, asking only for this: warmth on the skin, nicotine between their ruined teeth and in their ruined lungs. Who knows what old loves and conversations, what music sounded in their minds.

Brenda Walker

Brenda Walker is a writer and Professor Emerita of English and Literary Studies at the University of Western Australia. Her work has won numerous prizes, including the Victorian Premier's Award for Non-Fiction and an O. Henry Prize.

Waiting for a friend

KATE COLE-ADAMS

Early on the morning of my fifty-ninth birthday I carry a mug across the backyard and into the studio to wait for Harriet. We haven't been in the same country in nearly a decade, but through the alignment of datelines and digital technology we create a nest of impossible time: me in Melbourne (PJs, Ugg boots, celebratory cup of tea); she at her kitchen table in Devon, England, late evening on the second anniversary of the day her husband took himself to the highest point in the small town where they had loved each other for twenty years, and jumped.

I've known Harriet since I was eleven. She was friends with Jo whose twin sister was friends with me. We all lived around the corner from each other in Islington, London, where my father was posted for five years as a correspondent for Melbourne's *Age* newspaper. I'd seen her around and had a vague idea that she might be a *bit annoying*. Certainly, she was exuberant (the great wide smile; the sense of all of her bounding forward at once). My reserve lasted until we found we were enrolled in the same secondary school, at which point our parents arranged a get-together and we fell in love.

For the next three years we were each other's. Not exclusively, not in a romantic sense – although at her family holiday house in Norfolk we enacted long elaborate dramas in which she (as casting director) was inevitably a girl, I a boy, and in which I once lay on top of her in a field and we pressed our lips together, hard and ardent, until the air around us felt all empty and we stopped. We wrote each other poems.

This lasted until I was fourteen and we both left London, she for the pastel-tinged haze of Cambridge, me back to the blasted wastelands of suburban Melbourne 1975. Since then I've returned to England maybe half a dozen times. Between these visits our contact has been sporadic.

So here we are. Harriet is wearing long tasselled earrings that sway as she talks, and a black dress with what look like lurex thunderbolts. This is not in itself significant; she wears much the same at our usual mid-morning (her time) catch-ups. ('I walk across the moors like this; I don't give a shit about appropriate clothing.') It is 10 pm in her world – on deathday, as she calls it – and she is holding up a lighted candle in a jar depicting a Moomin sunset, which she rotates slowly as she turns off the lights and sings happy birthday. Then she tells me about her day.

In a world without coronavirus, we might never have had this conversation. We would have already seen each other in person earlier that month. April 2020. The trip booked and paid for. And after I got back to Australia, I may have called her, or she me; but probably not on Zoom, because – why would we? We'd never heard of it.

Harriet isn't the only English friend I've been Zooming. Before the lockdown, these encounters would have taken place, as with the impossible birthday/deathday conversation, uncomfortably early for me and awkwardly late for them. Now, we can talk quite easily on the same day, albeit me in wintry darkness; them in summer's light. We meet around 7 pm Melbourne time, when most of my friends would usually be working somewhere other than their kitchen tables, and therefore unavailable. Now they nurse mugs while I sip spiced rum (another COVID mutation) and ask them about then and now and what has happened in between. This knitting together of space and time creates its own new sorts of meaning.

On the floor in my study I have a red plastic folder until recently jammed with cards and envelopes and slips of blue folded paper labelled *Par avion*. Some are decorated. Some contain small talismans. Most are postmarked between 1975, when my family left London, and 1978, when I finished school in Melbourne. This time capsule has waited, barely touched for half a century, lugged from home to home, tucked into chests and boxes, the back of my mind. When I unpacked the contents a few weeks back, their physical presence, the crystalline residue that emanated from them, was so palpable I had to put them down and go for a walk.

It is almost impossible, I think, to communicate to someone who has grown up with an internet connection and access to cheap air travel the vastness of the expanses between continents. When 17,000 kilometres was exactly that. A letter sent from

Melbourne in 1976 could take a fortnight or more to reach London. Phone calls were for birthdays and Christmas: planned for weeks, measured in minutes, and what to say?

The letters now spread around the floor near my desk describe a world not only lost but, for me, never attained. ('*R is, at this moment in time, going out with three whole boys at once . . .*') In the year or so before leaving, I had been aware that friends (Harriet, notably) had grown breasts; some already had boyfriends. But the me that left London was still lanky and boy-chested; she didn't menstruate; she slept surrounded by photos of Paul Newman and the Beatles. She yearned in complicated, non-specific ways for boys to kiss her or like her or lie on top of her. She was not entirely naïve; she knew things. She knew, too, that she had to go back to Australia, but she was entirely unprepared for the realities of the return, when the part of her that had been moving forward would stall, and the people she loved would move on without her. And so, in a sense, would she.

'*I suppose you must have heard of the music scene in England. Punk rock. Of course.*'

Eventually I would make new friends; get a life, a career, partners; have children; write books. But it would take the pandemic to lead me back to my old friend, and my young self.

I've been thinking a lot lately about the nature of time, and what it is and how it passes and how it lodges itself in our bodies. In that first panicky week of lockdown, I sense it flickering through my chest like a queasy pulse. And in the weeks that follow, as the world contracts, I feel the enforced passivity, the curfews and constraints nudging me closer to childhood.

When the interviews I have scheduled for my (formerly) upcoming trip to London have to move to Zoom, I am doubtful. I fear our talks will be thinner, more detached. And yet I find that something happens in these conversations, cocooned in our individual studies and living rooms, mediated by the virus that separates and connects us. I am aware of a vulnerability (mine and theirs) that I attribute at least partly to the overwhelming fact of the pandemic; but also to the medium, which allows me to view in intimate detail my friends' expressions while I recede like an animated postage stamp to the top right corner of my screen. Because of the slight time lag, it is difficult to interject without disrupting the flow. But that small constraint intensifies the quality of attention I bring to the moment. I find that I see more, listen better.

I find that I have missed a lot of things along the way. The world for which I mourned so long was darker than I'd understood. One friend has described lying curled in bed blocking out the sounds of her parents' violent arguments. Others don't want to talk about those days at all.

The last time I saw H in person was nine years less three days ago, on the morning after the surprise fiftieth birthday her husband, Ed, had organised. As a young woman there'd been plenty of boyfriends. Then there was a husband with whom she had two sons. And then there was Ed. 'The first time we kissed was on November 25, 1998, which was the night we were doing a massive show for school. And I kissed him, and I said, "Oh *no*!"– because I knew then. Because he smelt right. And I knew ...'

Genial, kind, handsome Ed. Who taught us how to ring the bells at the local church; who guided my partner and son across the moors; who gave no sign of the noises inside his own head; or not that I could tell. And where was Harriet in all this? Everywhere. All at once. Did I notice how hard she was working to keep us all fed, entertained? Was her voice a little too bright, a little brittle?

In the weeks leading to and from the anniversary, our conversations will start tentatively (will there be enough to say?) in a semi-formal interview mode, and veer almost immediately and delightfully off track. I rediscover the rhythms and rich cadences of her voice, the mobile face, emphatic opinions; her generosity, curiosity, the uprush of her laugh. We talk about ageing and sex and body hair. Words and grammar. The declensions of grief. We talk about #MeToo and #BlackLivesMatter and the falling sky. We talk about Ed.

One of the hardest things about the process of grief is the way in which it repurposes your memories. Like the discovery of an affair, it orientates everything it touches: past, present, future. The first time Harriet and I speak by Zoom, she talks about the pain of recalling almost anything from the past two decades. 'Even the happy memories are painful. Because it's all being re-remembered.' She talks about being dragged *against my will, kicking and screaming* to this part of herself. The part that looks back.

We talk about that. And we talk about the stories we tell ourselves about love. The script our culture has written us and the parts that we play, and how seductive they are, and how we are seduced. And how it is kind of crap. But compelling crap.

And we talk about each other.

Until very recently if you had asked me about my London childhood, I could have described in detail the friends I'd left behind, and the desolation of my return. What I couldn't have told you was what I was like. Because I had no idea. Because when I left London, I left myself too. Or that's how it seemed. But Harriet remembers.

'To me it felt like we were adventurers together,' she writes. 'Carelessly and optimistically I would set out with no plan and definitely no figurative bottle of water in my figurative jungle, because I could rely on you to remember those things and in any case your plan, when it emerged, would be so much better than mine. You were Watson to my Holmes . . .'

And a little later: 'Funny, sensible, gentle, incandescent Kate, who made me feel mighty and strong because she had chosen me to be her best friend.'

She tells me too about the relief, through our conversations, of travelling back and finding a part of herself untouched not only by Ed and his insistent, ambiguous legacy ('my husband had an affair with death') but by the postures and pressures that came with puberty and the ambiguous gift of womanhood.

'And what do I feel about that girl now? I feel she's – I *like* her. I like her. And that's a big deal!'

Only once in the time we've been speaking have I seen her unable to laugh. This is the week after we talk at length about her falling in love with Ed. In a later conversation she will tell me the toll of that call: the lethargy, the sense of a day spent climbing stairs. 'You know, one of the things I find saddest about Ed going, which I finally admitted to myself, was he took with him a bit of me that I really like.' When I ask, she says, 'My softer side . . . loving, you know, the loving.' Pause. 'Yeah, that's something I thought I did quite well.'

And my heart squeezes slow sad blood for her, and I am reminded that these conversations do not exist out of time but create the conditions for what comes next.

But today she is ready. Over the next hour we will cover: our views on tattoos; her love of Dickens; the importance of reading trash; her delight in her four young-adult children, three of them now in lockdown with her; the letter of resignation she has just sent to the school where she and Ed met and where she was head of English and drama; the formative exotica of London's long-gone Biba boutique; the sex/death nexus ('The awfulness of your libido suddenly waking up at completely the wrong time when you're like, What? *What?*') The self-help book she is writing for families reconfigured by suicide.

We end in what now seems to be our normal: a flurry of two-handed waving, kisses. To me she is as vivid, beautiful, her smile as entire, as when we first met; unmarred by time, space, the passage of grief.

'See you next week.'

Kate Cole-Adams is a writer and journalist. Her 2017 book *Anaesthesia: The gift of oblivion and the mystery of consciousness* won the Mark and Evette Moran Nib Literary Award and was shortlisted for the 2018 Victorian Premier's Literary Award.

Too deadly: Coronavirus in Blak Australia

MELISSA LUCASHENKO

I'm Aboriginal, so this is where I am supposed to tell you exactly how awful COVID has been for us, as the global economy tanks and the national health system creaks under the strain of the pandemic. How 'we're all in this together' – but maybe some of us are more in it together than others.

Here's where I write about how the Indigenous poor have gotten poorer still, and the Indigenous sick more wretched. I suppose the dead of all races can't get any more dead, but in a country where us mob linger at the bottom of every statistical table, how could COVID be anything but yet more terrible, disabling news for the First Nations?

Well, here's the thing. When you are Indigenous, and when you live as colonised Indigenous people for generations, your mob learns certain things. About exclusion, for instance, and the meaning of marginalisation. About how to distinguish necessity from luxury, and truth from lies. About how to survive generation after generation of externally imposed hard times. I'm not saying we haven't been affected by the health crisis. First Nations mob are two to three times sicker than mainstream Australians. COVID hitting us, especially in remote areas, would have been bloody terrible. For some individuals, it was. My own white sista Deb Kilroy was one of the first people in our community to catch the virus, along with young activist Neta-Rie Mabo, who had flown with Deb on the same international flight. Goomeroi singer-songwriter Thelma Plum was also infected. All these sistas (immediately dubbed the 'Typhoid Murries') had to isolate for long and difficult weeks. Deb and Neta ended up in hospital in Meanjin, sick as dogs. So it's not like it didn't affect my Blak community. It's just that there's more to the Aboriginal story than suffering. There always is. For Indigenous people, COVID has been both easier and harder to live through than for others on this continent.

Let's start with the ridiculous: the toilet paper wars. Forget the virus – like the Muslim community, Aboriginal Australia was in grave danger of dying of laughter at the sight of mainstream suburbanites actually punching on over bog roll. My fb thread began to resemble the Monty Python skit – *Toilet paper?! Luxury!!* Blackfellas from Perth to Penrith began reminiscing about squares of newspaper hanging on a nail in the thunder box. Or using rags when homeless, or stripping paperbark trees out bush. I was simply agog: in a nation where nearly all whitefellas have access to *drinkable* running water, there was panic over people not being able to remove the shit from their bodies. Freud would have had a field day. Some blackfellas reckoned COVID was almost worth the stress, just to see dugais on the news, brawling in Coles over their precious twelve-packs of Sorbent. (Another silver lining: for once we weren't the only ones being profiled and stalked by store security.)

Aboriginal people and organisations swung into action fast all over Australia, determined to protect our precious Elders and vulnerable others. The mob solidarity born out of classical Aboriginal culture was there as it always is; Australia might only rarely value our lives, but we do, and we know how to stick together to protect them. As Blak journalist Amy McQuire rightly noted, the Aboriginal health networks didn't wait around for white governments to save us. Remote communities locked themselves down early, and hard. Most remain locked down now, in mid-2020, doing what needs to be done to shield themselves from a mainstream Australia even more fatal to black lives than usual. 'Sunshine doesn't kill the virus,' one Aboriginal health video advised, 'and rain won't wash it away.' A near-toothless Aunty with silver hair was roped in to deliver the message that COVID was *really really serious, you mob*. Materials were rapidly produced in multiple Indigenous languages for those with limited or no English. The faces in all these videos had to be blackfellas, and they had to be grassroots faces, because most Aboriginal people have very little reason to trust white authority.

My Asian friends were quick to don surgical masks. But some of my Aboriginal rellos were – and remain – highly sceptical about social distancing. Since when did Australian governments ever want to help blackfellas? What were they trying to gain by keeping us apart? If Blak togetherness and solidarity are our saviours – and they most definitely are – was social distancing a distraction from some other, more sinister agenda? Conspiracy theories abounded. African babies were being abducted and experimented on by Western scientists trying to develop a vaccine. The virus was deliberately introduced, by the Chinese, by the Russians, by the right wing here. Take your pick. Crazy talk, yes. And inexplicable to outsiders, until you remember that the government was injecting young Aboriginal women with Depo Provera a few short decades ago. Injecting our people without informed consent in an explicit attempt to prevent Aboriginal babies being born. It was also not lost on our mobs that the British invasion in 1788 brought smallpox to these shores. Nobody has yet been able to either prove or disprove that that virus was deliberately introduced to Sydney. Native Americans were purposely given pox-infected blankets, though, and it's plausible that the same was done here. Biological warfare doesn't feel new to us; it just feels like the logical extension of overpolicing Aboriginal suburbs, or defunding our Blak services, or telling us to 'have a go to get a go' when our imprisonment rates are higher than blacks in apartheid South Africa.

We all live in an era of failing Western 'democracies'. Mainstream citizens like my white working-class neighbours are also disaffected and grumble about 'elites' and 'Canberra bubbles'. But if you aren't Aboriginal, you will have no generational memory of Australian governments actively trying over two centuries to wipe your mob out. No knowledge that, to a large minority of fellow citizens, your culture, your life, you – are deemed utterly worthless. (Don't believe me? Try reading the comments under literally any Aboriginal article in the mainstream media.) Governments didn't have explicit policies of removing Aboriginal kids for over a century through benevolence. They did it to destroy us as Aboriginal. To make us European in genetics and in culture. That's why little Koori kids in NSW institutions were encouraged to pray that they'd wake up white. That's why so many of us have pale skin today. We have been the targets of state-sanctioned genocide since Australia began.

So our First Nations medical services and other Indigenous health workers had quite the job on their hands to convince blackfellas that COVID mattered, and it wasn't just government spin. To protect us from the mainstream inertia and the massive underfunding. It's a job they did very fast and very well. Along with the state border lockdowns and other necessary measures, Blak health experts and health workers saved many, many lives of the sick and the old. Those aren't lives you'll see celebrated often. But we know what they're worth, because Elders especially hold our memories. Of traditional culture, of traditional medicines and practices and languages. Of survival.

It's no accident that we have managed this pandemic as well as we have. Survival is what we do. First we made it through the Ice Age. Then the catastrophe of British invasion: pox, guns, all the rest. Poverty, capitalism, ongoing child removal. Policies of attempted genocide. Whatever history has thrown at blackfellas, we have survived. And often thrived.

When COVID forced Australia to stay home for long tedious weeks, the wailing about being 'locked up' and 'imprisoned' came loud and long from the middle class. *Uh-huh*. Being imprisoned, either literally or else metaphorically by poverty (getting further than the local park takes cash) is something most blackfellas have experienced. And it doesn't resemble sitting comfortably at home with Netflix and Uber Eats. So cry me a river, bitches. The iso most Australians have come through is laughably easy compared to being locked in a racist white institution with your psych meds abruptly ripped away, without family visits, without anyone who speaks your Aboriginal language, sans power or dignity or human rights. Try being a rape survivor repeatedly strip-searched behind bars by armed strangers before you complain about being 'imprisoned' in your suburban home.

When supermarket food seemed like it might run out, and Australia went berserk stockpiling, we reminded each other that we have lived off the land here for upwards of one hundred thousand years. And that plenty of our mob with access to land still do. Uncle Dennis Foley grew up eating bush tucker from the Narrabeen swamp and roadkill off the northern Sydney highway; he might be a Canberra university professor today, but the lessons from grassroots life are still the most important. *Look after country so country can look after us.* A Goorie neighbour simply nods when I talk about the Herefords near our place, and how we can knock a few on the head if the supply chain fails. 'Just common sense, ay, sis.' Maslow's hierarchy has never struck us as theoretical: food, water, shelter, mob – these are the essentials of life. You have to be pretty high up the white supremacist hierarchy to lose sight of that simple fact.

In my own family, we grew most of what we ate as kids, and it was a rare trip that didn't see Mum pulling the car over to collect free food: wild guavas, footpath mangoes and paddock mushrooms, all supplements to what the chooks laid and our own labour in the garden produced. Perhaps because I'm the child of refugees as well as being Aboriginal, I've never trusted the mainstream economy not to throw us to the wolves. Capitalism as we know it has consistently failed the poor along with the Blak. Not to mention the environment on which everything rests. It didn't take us COVID to realise that. Plenty of First Nations cupboards were already empty long before February 2020, and will be for a long time after the pandemic is gone. There are lots of regions in Australia, even in major cities, where it isn't safe to tread while visibly Blak, because of vicious skinheads and murderous vigilantes and the ever-present reality of police who kill.

In May the pandemic in Australia began to temporarily ease. The island continent was largely protected by geography and by sustained community effort. Then footage of a white cop casually kneeling on the neck of a dying George Floyd for nearly nine minutes galvanised the world. Blackfellas and our allies here in Australia came together, along with African-Americans, the Black British, populations in countries everywhere that white supremacy costs lives. We gathered by the thousands, by the tens of thousands. Our rage and anguish were palpable.

I caught a bus into central Brisbane with a couple of dozen other Aboriginal mob bearing signs, wearing the colours, determined to bear witness to Floyd's death, which stood that day for all our deaths at white hands. The prime minister warned that our protests were dangerous, anti-social; Senator Mathias Cormann later went on to say they were 'self-indulgent' – pretty rich, I thought, coming from the man who shook hands with Fraser Anning immediately after Anning deliberately uttered the words 'final solution' in Australia's parliament.

The bleating of these politicians was a perfect demonstration of just how little our Black lives matter to authorities. Dog-whistling about demonstrations while supermarkets opened, as malls were full of shoppers and schools accepted pupils back, was transparently not about public safety. Indeed, when the numbers did spike again we were handed proof on a plate that outdoor demonstrations were not responsible for spreading the contagion. It was about state control, and government spin, and slandering blackfellas as we have always been slandered, for being dirty, diseased and dangerous to good white folk. We knew this, and gathered anyway, knowing there was a small but significant risk from COVID; a risk that was far outweighed by the terrible ongoing cost of white supremacy and murder by cop.

I got off the bus and made my way towards the front of the massive crowd with a poem I wrote after the killing in custody of Kumanjayi Yock in 1994. When I got near the podium – it took nearly twenty minutes – I realised that the immediate families of the dead were speaking, and stood back. All around me were faces – African, Asian, Muslim and Pasifika – every one covered with a surgical mask. Lots of white supporters and more blackfellas than I've ever seen in one spot. Maori brothers and sistas did a haka for us, and made us cry. Each of those people at the rally had taken a personal risk to come out. And everyone there clearly understood that while COVID is *really really serious, you mob,* the virus of racism is even more dangerous to Aboriginal bodies. And nobody is putting millions into a vaccine for racism. Racial capitalism is far too profitable.

Among the tens of thousands rallying in King George Square, I saw a familiar face under a white surgical mask and embraced its owner. After surviving the COVID that nearly killed her, Deb Kilroy looked utterly exhausted, with a lined face and red, swollen eyes. At the time of writing my sista has been COVID-positive (but mostly non-infectious) for more than one hundred days. Neta-Rie Mabo and Thelma Plum were also nearby in the throng. COVID is devastating, we all know that. But then so is everyday life in Australia if you look Blak, or African, or Muslim. We don't get to put our face masks on lightly; we have to do it knowing that those same life-saving masks can be deliberately misinterpreted by racist cops, and used as excuses to arrest, bash or kill us.

So do this one thing for me. Before you listen to the Morrisons and the Cormanns or their various mates who have made blame-the-victim the Australian national sport, use your imagination. Imagine the trauma of discrimination, imagine the sheer terror, which made us mob rally that day anyway, knowing how many hundreds of thousands have died from COVID. Ask yourself why we took any risk at all to say something as simple as Black Lives Matter.

Imagine if that cop had listened to the shouting and risen to his feet, and taken his knee off George Floyd's neck in time for Floyd to gulp a big, lifesaving breath of air.

Imagine if Kumanjayi Walker got to stay asleep on his lounge-room mattress.

Imagine if Ms Dhu had been helped.

Imagine if Tanya Day had been put into a taxi, not a coffin.

Imagine if blackfellas didn't always live with the terror of imminent death hanging over us, not just during this pandemic, but every damn day of every damn year.

Imagine if all lives mattered.

Melissa Lucashenko is a First Nations writer of Goorie (Aboriginal) and European descent. Her novel *Too Much Lip* won the Miles Franklin Award in 2019.

Trouble breathing

JENNIFER MILLS

'Practice respiratory etiquette'
World Health Organization advice for COVID-19

The medical term is Acute Respiratory Distress Syndrome. On an X-ray, the black space of a healthy lung fills with a cloud of white. Under a microscope, the delicate walls of the alveoli thicken. Oxygen from the breath must pass through these walls and into the capillaries to enter the blood. In a patient with severe COVID-19, oxygen intake is drastically limited. Hypoxia rapidly develops into systemic harm, and, in around half the cases that require intubation, death. It is traumatic, both for the person dying, and for anyone watching as they gasp for air.

Nurses, doctors, cleaners and other staff in Italy's hospitals took great risks, working long hours in crisis conditions with limited PPE. As COVID-19 overwhelmed some regions' intensive-care capacities, momentous decisions about life support had to be made. For those on the front line, the virus was sometimes a direct question of who should get the help they needed to breathe.

The breath is an autonomic function of a healthy body; you don't have to think about doing it. In distress – a smoke-filled environment, a panic attack, or choking – you suddenly become aware of the body's dependence on air. You can survive for a few weeks without food, a few days without water. Without oxygen, you have a few minutes.

In the early days of the pandemic here in Turin, I started to visualise the breath in a new way. As I stepped into the middle of the road to keep my distance from someone, or stood a metre away from a person I was in conversation with, I realised that I saw my own breath, and the breaths of others, as clouds of vapour. I imagined airborne droplets around us, filled with some of the microbial life forms – bacteria, viruses – that lived in, and formed, our bodies. We were all in flux, in constant exchange with each other and with the atmosphere. I'd known this before, but now I felt it: the human body was not an integral unit, but part of a system. We were not just individual actors, but habitats, hosts; bit players in a complex ecology.

Unlike other autonomic functions, the breath can be consciously controlled. I found that I was breathing more shallowly in public. I flinched at other people's coughs, and repressed my own. I wasn't too concerned for my own health, but I had begun to see myself as a vector of illness, a potential spreader. The idea that my exhalations might be lethal to others revealed a moral entanglement. That simple act of giving and receiving air that I had been doing all my life, mostly without thinking, could put strangers at risk.

Italy's stay-at-home orders seemed extreme, before they went global. We watched as other countries tried to dodge their own lockdowns, fearful of the costs. It was true that several forms of the unnecessarily multi-layered Italian police forces prowled the streets, and that people were being fined for violating the orders. But Italy was also the first Western democracy to institute lockdown measures, and it would have been impossible without the consent of the people. Our lockdown came with the slogan '*Io resto a casa*': I stay at home. The use of the first person invoked a sense of responsibility that was essential to its success.

Moral entanglement quickly developed into moral action: here was something we could do. The seemingly ethically distant act of staying home could reduce the opportunities for transmission and help to save lives. This small act of collective power made other actions possible.

The social technology of solidarity is nothing new, but in the first half of 2020 it has felt radically necessary. It was happening locally: in my *quartiere*, 'solidarity boxes' appeared on the streets and were filled with food for those in need. Neighbours checked on each other. Flash mobs and music kept our spirits up. Children made banners and hung them from balconies, proclaiming that all would be well. But it was also happening online and internationally, as we shared care and information, art and recipes, tech support and jokes; as we passed on the news, and learned one another's survival strategies. We were physically separated, but we were not isolated. We had all become agents in a network of care.

Italy often seems a divided country, where corruption is rife and trust in the state is low. As a foreigner, I was amazed at how quickly this sense of collective agency was mobilised. Perhaps it was because elders are valued here, or because the analogy of war evoked an intergenerational debt – many of those dying had survived the Second World War; some had fought fascism. But those same networks had mobilised in Australia during the bushfires, and were mobilising again as the pandemic response grew serious. Something similar had happened in Wuhan: alongside, and often in spite of, an authoritarian surveillance state, the collective sensibility ran deeply within Chinese culture.

It wasn't trust in the state that mobilised people, but trust in each other.

These networks of solidarity and mutual care are similar to the breath, in a way. They're almost an autonomic function. In a healthy society, a strong community, you don't have to think about them most of the time. But when you are struggling or ill, they are life or death.

In late May, as Italy began to emerge from lockdown, a video of the killing of George Floyd reignited the Black Lives Matter movement. Floyd's dying words, 'I can't breathe', were already a familiar refrain. They'd been said eleven times to New York police officers by Eric Garner in 2014. They'd been said twelve times by David Dungay Jr before he died after being restrained by correctional officers at Long Bay in 2015.

We all breathe the same air, but we don't breathe equally. As the Black Lives Matter movement returns with fresh vigour, it arrives with the lessons of the pandemic in hand. There are clear links between the direct state violence experienced by George Floyd and the indirect state violence which distributes health care unequally, which removes the safety nets from under certain communities, and which bails out businesses and leaves ordinary people to fend for themselves.

In the USA, the links were brutally clear. Black people were dying from the virus at twice the general rate; when the protests began, one in two thousand Black lives in America had already been lost to COVID-19. Writing in *ProPublica*, journalist Akilah Johnson pointed out that the pandemic 'has laid bare the structural racism baked into the American health system'. At the same time, shared vulnerability has helped bring people to action. One Black Lives Matter founder, Opal Tometi, told the *New Yorker* that the pandemic has made people 'more tender or sensitive to what is going on'.

'I can't breathe' was spelled out at the Black Lives Matter protest in Meanjin (Brisbane) on 3 June with 433 candles: 432 for the Aboriginal people who had died in custody since the Royal Commission ended in 1991, and one for George Floyd. When asked about the protests, Prime Minister Morrison said 'we don't need to draw equivalence here'. But to many, the patterns are obvious. Organiser Bo Spearim told NITV, 'Aboriginal people, black people, Indigenous people, we've made that connection.'

The pandemic has highlighted gaping inequalities in Italian society, too. *La Repubblica* estimates that half the Italian workforce has applied for emergency payments; the state's capacity to provide was quickly overwhelmed. According to Istat, Italy's national statistics agency, 27 per cent of the Italian population is at risk of poverty. In the south it is 45 per cent. With no savings to get through quarantine, the poorest have suffered most, particularly migrants. A million additional people have accessed food aid.

Italy has allocated temporary residency to over half a million undocumented farm workers, and there is a renewed campaign to pass *jus soli* laws that have been on ice since 2017, laws that would allow citizenship to anyone born in Italy. Black Lives Matter protests here are connecting police violence and the deaths of migrants at sea: one placard read, 'Under water, under the knee, I can't breathe.'

'"I can't breathe" suddenly equates racism with the deprivation of air,' wrote Ben Okri in the *Guardian*. '"I can't breathe" hints at the apocalypse of human values.' The question of who breathes, and who suffocates, is a question of who deserves to live, of who is human. A question of who is part of a community of care, and who is exiled from it. A question that will only become more urgent as the climate crisis develops.

In the first weeks of Italy's lockdown, Australia appeared to exist in a previous version of reality. It seemed a fragile place that couldn't see its own fragility, a place where a fiction of innocence still prevailed. People went about their ordinary lives, perhaps reconsidering their travel plans. The prime minister announced a ban on large gatherings, then attended a 3000-person church service before it came into effect. I worried that our leaders weren't taking this emergency seriously.

I often write about the climate crisis, so I've had some version of this Cassandra complex for a while now. Long fascinated by the particularly Australian intersection of climate denialism, colonisation and selective amnesia, I was afraid that all these familiar mechanisms would lead to catastrophe. It wasn't unreasonable to expect disaster; I'd already watched it play out once this year, during the worst bushfire season on record. The scale of those fires would once have been unimaginable, a scene from climate fiction. Now the media was calling it a 'new normal'.

But normality had already left us. The climate emergency was moving fast, and it was getting personal. While ordinary people were quick to act in solidarity and care, finding each other food and housing or raising much-needed funds for our largely volunteer rural firefighters, the federal government had been just as quick to pass responsibility to the states. I read the letter that a group of senior firefighters had sent the prime minister earlier in 2019, warning of disaster. I saw the photos of him on holiday in Hawaii. I've been a volunteer firefighter for eight years; I know how risky it is. And there these politicians were, gaslighting the public on the reality of the climate crisis, even as people were dying from it.

Friends that had bought P2 masks to get through the fires were using the same masks to get through the pandemic. Both events involved a struggle for air, an awareness of the breath that is at once symbolic and deeply, physically real. Taken into the body, I hoped this shared vulnerability might become transformative.

As Italy rebuilds, politicians here speak of constructing a greener, more egalitarian country. The EU is placing the European Green Deal at the centre of its recovery, doubling its Just Transition Fund and allocating billions to climate action. I'm struck again by the sense that Australia exists in another version of reality.

Spared the worst of the pandemic so far, its possibilities for change are already being snatched away: free childcare replaced by gas fields, income support evaporating, dreams of shiny new kitchens in unaffordable houses. Our government handed the gas lobby the keys to the family car, in the form of its COVID-19 co-ordination commission, and the results have been depressingly predictable.

The climate crisis is a public health emergency. If it moves more slowly than the virus, it's also far more lethal. Like the pandemic, its harms are unevenly distributed, exacerbating pre-existing inequalities. It demands urgent action, global co-operation, scientific and medical expertise. It is inseparable from social transformation, from an awareness of how much we all depend on each other – and how much our collective power can achieve. In a just world, no one should have to fight for oxygen.

Jennifer Mills is an author, editor and critic. Her novel *Dyschronia* was shortlisted for the Miles Franklin Literary Award in 2019. *The Airways*, a queer ghost story set in Sydney and Beijing, will be published in 2021. Mills lives in Turin, Italy.

A long shadow

GEORGE MEGALOGENIS

The coronavirus launched its silent assault on the Australia Day long weekend without revealing its ultimate weapon. This placed the community in far greater danger than was appreciated at the time.

The first group of cases, announced on 25 January 2020, were treated with appropriate alarm by the then Chief Medical Officer, Dr Brendan Murphy. But the initial advice he provided to the public gravely underestimated the threat. Dr Murphy believed the virus could be fought without disrupting national life. He saw no need to close the border to prevent its spread. For those returning travellers who had been in the epicentre of the outbreak in China, he recommended that they see their doctor, or go to an emergency department 'if they become unwell'. Dr Murphy's advice gave the virus a head start because it didn't wait for a patient to show symptoms. It was Australia's good fortune that we learned of our mistake just three days later.

On 28 January, Australian officials received word from researchers in Germany that the virus could be spread without the carrier knowing they had it. A local businessman had been infected by a visiting colleague from China who seemed perfectly healthy – there was no coughing, sneezing, fatigue or fever to alert either person. A similar case was reported in Japan.

Health officials the world over had assumed that this new virus would behave like its more deadly genetic cousin SARS, which jumped from person to person only once the carrier showed symptoms. But a silent transmission was a game-changer. It meant hundreds, potentially thousands, could contract the coronavirus before authorities knew what had hit them. To combat a pathogen this insidious and infectious, nations would need to shut down much of their economic and cultural life. They would have to close their borders to foreigners, place returning travellers into quarantine for fourteen days, and aggressively test their local population.

Australia had two advantages in this fight. One was natural: we had a continent to ourselves. The other was man-made: our public health system.

But these advantages could be undermined by complacent policy making, as the United Kingdom demonstrated. The UK was physically separated from the continent of Europe, and it had the National Health Service, an institution that was founded in 1948, thirty-six years before our own Medicare. Yet neither advantage was exploited. The border was left open into May, more than a month after everyone else had shut theirs. Meanwhile, the NHS had been run down by a decade of austerity, and was unable to cope with the caseload. At the time of writing, the UK had the second-highest death rate in the rich world after Belgium. This perspective is important to bear in mind. Australia made its own luck. The key was a willingness to learn from early mistakes.

Initially, Dr Murphy, along with his fellow state and territory chief health officers at the Australian Health Protection Principal Committee, thought the virus could be managed by self-isolating from just one country, China. The decision to close the border to our largest trading partner and second largest source of new migrants after India was taken on 1 February. The critical piece of information for Dr Murphy was the evidence of infections spreading beyond the Hubei province, in central China.

The most curious aspect of the policy response was not the country-specific ban, but the political theatrics associated with its execution. China had already locked down the city of Wuhan, where the virus was believed to have originated, as well as sixteen neighbouring cities in Hubei province. That meant we wouldn't be stopping many Chinese citizens from coming here, because they were already being prevented from leaving. Of more interest was what would happen to the Australian citizens and permanent residents who wanted to return home from Wuhan. We wanted to let them in, and the Chinese gave us a small window to move before they stopped all travel. In the end, the two countries negotiated the evacuation of more than 500 men, women and children.

The first planeload of 243 left China on 3 February. Their destination was Christmas Island, where they would remain for a fourteen-day quarantine before returning to their lives in

Australia. They were joined by another thirty-six evacuees, who were part of a New Zealand-operated airlift.

The deployment of Christmas Island was problematic. This particular offshore detention facility was the desolate symbol of Australia's border protection regime. Now it was being enlisted to suppress a virus. Whether it meant to or not, the Morrison government risked equating the evacuees with the asylum seekers that had previously been detained and demonised on Christmas Island.

A second flight carrying 266 people left Wuhan on 9 February. But Christmas Island was already full, so the evacuees were taken to a disused work camp outside Darwin. Australians evacuated from the *Diamond Princess* cruise ship stranded in Japan were also brought to the mainland.

To be fair, these improvised measures were effective in containing the virus in February when it was already running unchecked in many other rich countries. Part of the explanation for the low infection numbers at the time was that Chinese Australians did not bring the virus with them. In fact, there were no cases at all among the 243 on Christmas Island.

But the message was lost on the bigots in the community. Complaints of racial abuse lodged with the Australian Human Rights Commission surged in February and March. One in three complaints in February alone were explicitly related to the coronavirus. Scott Morrison became increasingly frustrated with the attacks, and gave an interview to SBS in April to call on all Australians to unite. By choosing the ethnic broadcaster to deliver this message, he made clear whose side he was on. 'I am massively disappointed because it's just so wrong,' he said.

'It's so against who we are as a people, and it was the Chinese Australian community that protected Australia so early on in this virus outbreak around the world.'

The paradox, of course, is that Morrison sent them to Christmas Island in the first place. Perhaps he regretted the overkill, because he didn't repeat the exercise for people coming back from other virus hotspots. The border was closed to Iran on 1 March, South Korea four days later, and then Italy on the eleventh. Australians returning from these countries were told to self-isolate in their own homes for fourteen days; the same requirement that already applied to those who had made their own way back from China. But there was no enforced quarantine. Dr Murphy cites this as his one great regret.

'I would like to have formally hotel-quarantined people a little earlier, because most of our cases at that time were coming from overseas travellers, but there wasn't the room in hotels to do it,' he told the Senate Select Committee on COVID-19 in May. 'I don't think it would have been practical. So we did it as soon as we could.'

It was a catch-22. The hotels had to be put out of business by the closure of the border to all foreigners before there were enough rooms available to quarantine returning Australians.

The main threat to Australia at the time was from the North Atlantic, not Asia. The United States in particular had been slow to test for the virus, and President Donald Trump had dismissed the whole thing as a hoax. The privatised American health system was also woefully unprepared.

The evidence that the virus was running unchecked throughout the US was hiding in plain sight here. Holidaymakers returning from the Aspen ski fields had spread the infection across some of Melbourne's wealthiest suburbs in February and March. Then came the double shock of the Hollywood superstar and the Australian politician who brought the virus to Australia. On 12 March, the actor Tom Hanks and his wife Rita Wilson went into quarantine on the Gold Coast after they had tested positive for the coronavirus. The couple were in Australia filming Baz Luhrmann's biopic on Elvis Presley. The following day, as the first of the domestic restrictions were announced by the national cabinet of Commonwealth, state and territory leaders, Home Affairs minister Peter Dutton revealed that he had picked up the virus on an official trip to the US. Yet there was never any serious consideration given to imposing a travel ban on the world's richest nation.

This undue deference to the US meant that infections surged from fewer than 200 cases on 13 March to more than 2000 by the twenty-fifth. Two thirds were from overseas. The numbers were supercharged by the case of the *Ruby Princess*, the boat no one thought to stop, or at least test for coronavirus, before its 2700 passengers were allowed to disembark in Sydney on 19 March.

Dr Murphy offered a possible explanation for that infamous lapse in his evidence to the Senate Select Committee on COVID-19.

'It had only been to New Zealand and back, and New Zealand was not seen as a high-risk country,' he said in response to a question from Senator Kristina Keneally, a member of the Senate committee.

'I think everyone was quite surprised at the fact that there turned out to have been a significant COVID outbreak on that ship. On first principles it wouldn't normally have been seen to be a particularly high-risk vessel.'

Keneally: 'We certainly did know, though, that cruise ships presented a particularly significant risk. We'd had the *Diamond Princess* a month earlier.'

Murphy: 'Exactly.'

It appears that an unconscious cultural bias undermined our early response. We fell for the trap of treating this as an Asian virus, an Iranian virus, then an Italian virus. The global nature of the pandemic was not fully appreciated by Australian health officials until mid-March, six weeks after the first asymptomatic case was reported in Germany.

The border was finally closed to all foreigners on 20 March, the day after the *Ruby Princess* docked. By the twenty-fifth, Australians were banned from leaving the country and every state except NSW and Victoria was closed to interstate visitors. A fourteen-day hotel quarantine for all Australians, regardless of the country they flew in from, was belatedly introduced on 29 March.

These measures helped to suppress the virus in April and May, and moved Australia on to the shortlist of role model countries that had flattened the infection curve. Then a spike in community transmission in June, linked to poor hygiene and lax controls in a handful of Melbourne's quarantine hotels, provided a timely warning that the virus had not finished punishing complacency.

The pandemic has not only ended Australia's record run of economic growth, it has taken the recovery out of our hands while we wait for a vaccine or an effective treatment.

Australia's wealth relies on open borders. We earn export income from mining, agriculture, international students and tourism. That last part of the equation – migration – has been responsible for the majority of Australia's population growth since the early years of the twenty-first century. Before the lockdown, only two capitals – Brisbane and Hobart – and one region – Queensland – defied this trend by sourcing a majority of their population growth locally, through net natural increase and internal migration. In Perth, international migrants accounted for 53 per cent of the city's population growth, in Melbourne the figure was 66 per cent; in Sydney it was 85 per cent and in Adelaide it was 100 per cent.

There is no easy path back to prosperity as long as the borders remain closed or tightly controlled. This seems perverse on one level. Australia's reputation will be enhanced by its management of the health crisis. But while the virus is still burning around the globe, anyone who wants to migrate to Australia will have to be quarantined on arrival for at least fourteen days. They may face a second quarantine if they want to settle outside a capital city.

Australia has held the door open to mass migration since the end of the Second World War. Wave after wave, from Britain, Europe, Asia, the Pacific, the Middle East and Africa has transformed our identity. No other country we compare ourselves to is as diverse. We are a majority migrant nation, in which more than half the population was either born overseas or has at least one migrant parent. We crossed this threshold in 2018, when Australia's population hit 25 million.

Even a year without mass migration risks turning a

health-induced recession into an economic catastrophe. Sydney's property bubble could burst – perhaps Melbourne's as well. Adelaide's population will stagnate, and possibly start to fall. Large parts of regional Australia will face the twin pincer of population decline and accelerated ageing.

The examples of past booms and busts contain a recurring theme. The three long runs of prosperity in Australia's history – the wool and gold rushes of the nineteenth century, the post-war reconstruction of the 1950s and '60s, and our uninterrupted growth between 1991 and 2019 – had a dominant Melbourne in common. The Victorian capital was the fastest growing city in each boom because of its ability to attract and retain people from overseas, and to draw people from other parts of Australia.

Prosperity was shared in the closed economy of the 1950s and '60s. Adelaide, for instance, was lifted by the same migration and manufacturing waves as Melbourne. But in the open economies of the nineteenth century and again in the years since the reforms of the 1980s and '90s, wealth and power has been concentrated in Melbourne and Sydney.

Typically, Melbourne – the city that flew highest during the boom – suffered the hardest landing in the crash. The two most spectacular chapters were the long depression of the 1890s, and the deep recession of 1990–91. On both occasions, large numbers of people fled the cosmopolitan south for the frontier states – to the Western Australian goldfields in the 1890s and to early retirement in Queensland's south-east in the 1990s. This allowed Sydney to resume its position as the nation's unofficial capital, while Melbourne waited sullenly for the next migration wave to break in its favour.

Sydney is more vulnerable this time because its property market has further to fall than Melbourne's. Sydney was already losing 25,000 people per year to other parts of Australia before the first lockdown in March. Most were moving to Brisbane, where housing was cheaper. Melbourne, on the other hand, had been attracting people from other parts of the country, most notably Adelaide. But that internal advantage was lost once its five million people were forced into a second lockdown in July, and every state and territory shut out travellers from Victoria.

Yet Sydney and Melbourne are also the capitals most likely to find new drivers of growth while our door remains closed to the world, based on the talent they have already hoarded. Between them, the two cities account for just over 40 per cent of the nation's total population, but more than half of all its migrants, including more than 60 per cent of the Indian born living in Australia and more than 70 per cent of the Chinese born.

The virus will cast a long shadow, even after it is eradicated. The challenge for Australia is to maintain cohesion during the health and economic crisis. Otherwise, the recovery may find the nation permanently divided.

George Megalogenis is an author and journalist with more than three decades' experience in the media, including eleven years in the federal parliamentary press gallery. He is the author of five books, including *The Australian Moment*, which won the 2013 Prime Minister's Literary Award for Non-fiction and the 2012 Walkley Award for Non-fiction, and formed the basis for his three-part ABC documentary series *Making Australia Great*.

The ghost in the machine

JESS HILL

I can't put my finger on exactly when I first learned *I'm not in control*. My certainty broke gradually: the first crack when my body went rigid with a seizure and I convulsed violently into nothingness on a red-eye flight to Beirut; another splintering when the doctors found a 'mass' in my brain that afternoon; the edifice barely holding as my partner and I clung to each other all night in a skinny hospital bed, waiting for a prognosis of life or death.

Even then, as the sun rose the following morning and tears streamed down my face, the terror was strangely numinous, almost thrilling. This would be an aberration, a story to spice up a dinner party. My lifelong sense of certainty — a deeply habituated need for order and control — was fractured, but still intact.

But a few days later, the seizures began to roll. As soon as I closed my eyes to sleep, my eyebrow would twitch, and my right eyeball would pinball rhythmically in its socket. I would *feel* my brain untether from my spinal cord and float into an unknowable black.

During those nocturnal seizures, when I'd try and fail to *will* my brain back into place, I learned what it was to be out of control. Not just in that moment, and not just in my body, but existentially. In that little apartment in Beirut, I began to realise just how delusional the notion of control actually was.

But gosh it's nice to have an existential crisis when the world around you is basically sane. Back in 2012, Obama was in the White House, nobody had heard of ISIS, and Lebanon – a country I'd made my home – was defying the doomsayers by remaining peaceful. In Australia, the Gillard government had introduced the carbon tax, and it looked like finally we would incentivise industries to slash carbon emissions.

The big cracks in *that* charade – the one that had many of us believing the world ran on a badly flawed but at least fathom-able logic – appeared in 2016, as suddenly as my first seizure. Hours after American women lined up to place 'I Voted' stickers at the grave of legendary suffragist Susan B. Anthony, Donald Trump was elected, and instead of celebrating America's first female president, I was surrounded by young white men in MAGA hats at Sydney's most prestigious university cheering triumphantly, 'Grab 'em by the pussy!' There was the shock of Brexit – another crack. Syria's president gassing his own citizens with impunity – crack. Diplomatic silence on the internment of Chinese Muslims – crack.

But the edifice of public control didn't truly shatter for me until New Year's Eve 2019, when thousands of terrified adults and kids huddled under a dark red sky on the beach at

Mallacoota in coastal Victoria. 'It's fucking chaos,' said a man in ski goggles sitting on a boat just offshore. 'I've never seen anything like it.' For people without a boat, there was nowhere to go – more than 4000 people were trapped between the ocean and a gigantic encroaching fire, engorged by 80-kilometre winds. It was 49 degrees at 8 am. As the sirens sounded, they prepared to get in the water.

I wasn't on the beach that day. Days before, my family and I had left Lake Conjola as smoke blotted out the sky, and headed south along a Princes Highway bisected by charred and smouldering trees to Bega for Christmas with our relatives. A week later we fled back to Sydney, escaping with our two-year-old over Brown Mountain hours before the road out was closed. As the news came in about Mallacoota, our family in Bega was sheltering friends whose houses were expected to burn. I was scared for them – for all of us. It was then that I felt something inside me break apart. The terror of that scene on the beach – set against the utter intransigence of our on-holidays prime minister – smashed that already fractured edifice of order for me. Huge tracts of country unstoppably ablaze. Human control would not – *could not* – reassert itself. The people trapped on the Mallacoota foreshore were saved that day by a change in the wind.

Standing on the headland at Coogee Beach that New Year's Eve, I did my best to mirror my little girl's delight at her first sight of fireworks. But really I just wanted to run away. To disappear. To forget the sight of kangaroos caught in fences and to stop thinking about the billion animals killed by fire. To wait until Trump was voted out. To wait until climate change was magically solved. To take my little family somewhere far away, where the skies weren't choked with smoke. To hide until it was all over.

But there's no running from it. 'It' won't *ever* be over, because 'it' is not just one thing. *It* is not the catastrophic fires of last summer (or the ones yet to come), *it* is not climate change, *it* is not racism and police brutality, *it* is not the ongoing epidemic of domestic abuse and sexual assault, *it* is not Trump or Brexit or Bolsonaro, and *it* is not the coronavirus – though all are emblematic of the mess we're in. None of them are *it* – they are all branches from the same diseased tree, and the real problem is in the roots. The roots are old – dating back around 12,000 years – but what grew out of them is not an inevitable feature of human evolution; in fact, it's actually threatening our survival as a species. 'It' is, in short, the shift we made from societies built on the principle of balance to what we have now: a dominant culture obsessed with power-over and control.

There's a neat little term for this culture of power-over: patriarchy. It's a word that used to make people wince, but since Trump and #MeToo, it's been invoked with increasing regularity to explain the mess we're in. Its underlying principles are control and separation (which are romanticised as 'autonomy' and 'independence'). Patriarchy positions *all* people on a scale of entitlement to power and control: men have power over women, some men have power over other men, white people have power over people of colour, heterosexuals have power over LGBTQI, rich have power over poor, adults have power over children, all people have power over nature, and so on. Within this system, it is not individual men who have the most value, but men (and some women) who embody patriarchal traits of maleness: control, logic, strength, competitiveness, decisiveness, rationality, autonomy, self-sufficiency, heterosexuality (and − critically - whiteness). Men who don't embody these traits are assigned less value, and may be persecuted, attacked and shamed. That is how patriarchy polices men's allegiance: through shame, violence and fear.

We see the apotheosis of patriarchal behaviour in the most dangerous form of domestic abuse: coercive control. There's a moment when a perpetrator's campaign to assert power over his victim reaches a point of apparent success: his victim has come to understand that open resistance is futile, that her survival now depends on managing the perpetrator's need to feel in total control. Survivors talk of splitting into the role they play to survive − often to protect children as well as themselves − and the part of them they keep hidden, the place where they maintain their dignity. The perpetrator may allow this, or he may press into the furthest reaches of her being to snuff out any trace of her dignity. Here is patriarchy taken to its warped extreme: a system of control so committed that it is prepared to choke off its own need for connection rather than allow a shred of territory outside its control. This is why it's so hard for regular people to understand domestic abuse − it's deeply unnatural.

We are born longing for connection, for tenderness; to love and be loved. Patriarchy seeks to override those natural feelings in boys − to literally sever their capacities for emotional connection − by rendering those feelings weak and shameful. Even parents who consciously raise their sons to be compassionate and tender may see their boys shrink those parts of themselves in the face of social pressure and develop contempt

167

for traits considered 'female'. The reward boys are promised in exchange for betraying themselves is something else they learn when they are young: that they are entitled to power. But even those who benefit greatly from the power and privilege bestowed on them do not emerge unscathed.

If you have been socialised to believe patriarchal hierarchies are natural, you may feel entitled to the special privileges they grant you. You may also feel entitled to subjugate and harm what is positioned beneath you. Most people won't think of it as entitlement, or privilege; it's just what's necessary to protect your interests. This is as true of men who feel entitled to coerce and degrade their partners as it is of a white woman calling the cops on a black birdwatcher who asked her to leash her dog as it is of political leaders who install oppressive systems of surveillance.

Heading into the summer of 2019, Australians couldn't have imagined the deep and painful lesson we were about to receive on the limits of our patriarchal system, our obsession with control, and nature's indifference to it. In just a few months we lurched from the explosive horror of unstoppable infernos to pause in the relative relief of flooding along the eastern seaboard before emerging into what is possibly the most diabolical foe for anyone who values power and control: an invisible and untreatable pathogen creeping across borders, into our cities, into our homes, and, most treacherously, into our imaginations. It's little wonder we've seen the digital cult of conspiracy theories expand so rapidly during COVID-19, and that this has been weaponised by entities that profit from chaos and distrust. I can understand the appeal; it must be comforting to believe this pandemic is a fake and 'evil' ploy perpetuated by elites for global domination rather than a phenomenon that is beyond *anyone's* control.

The response to COVID-19 has shown us we can make huge changes, politically and personally. How that change looks – whether it is bent towards greater authority or greater equality – is up for grabs right now.

Can we imagine replacing a system of power-over with power sharing? Or are we going to double down on patriarchy and march obediently towards environmental collapse, towards a state surveillance culture designed to keep its 'enemies' at bay, towards deeper impoverishment for the masses while the super-rich hoard their money? Having lost control, do we really want to go back to how we were? If we don't change now, then when?

During this uneasy pause – as an untreatable pathogen forces us to shelter in place – we actually have a chance to look beyond the hour-to-hour dramas that consume us. From one perspective, we can throw up our hands and say everything's getting worse: more corrupt, more autocratic, more ignorant, more dangerous. That's all true, but there's another perspective that's just as relevant: we are living through an accelerated period of resistance. From the dogged exposure of paedophile protection rackets within the Catholic Church to Occupy Wall Street to the Arab Spring to #MeToo to the student climate strikers and Black Lives Matter, we are seeing enormous people-powered movements resisting and overturning the paradigms of power-over, of subjugation - of patriarchy. The needle is moving fast: a few years ago, it was unimaginable that we would be having mainstream conversations about patriarchy, or that 'radical' ideas like redirecting funds from the police to the community – a core demand of the Black Lives Matter movement – would be seriously discussed on our public broadcaster. Who could have predicted, before #MeToo, that we'd see one of the country's most esteemed judges – a position long considered untouchable – make headline news for his predatory behaviour? This is fast and furious change.

But this project to undo patriarchy cannot truly accelerate in Australia until we have an honest reckoning with our colonial past and present. My ancestors introduced a particularly virulent strain of patriarchy to Australia, supercharged with misogyny and racism, that tried to destroy cultures that had survived and thrived for at least 60,000 years on the exact opposite principles of patriarchy: principles of balance, of sustainability – of *power-with*, not power-over. These phenomenally successful cultures – in which men and women were emotionally embodied and interdependent, children were raised gently, laws were not bent to suit the powerful, and care for Country was indistinguishable from care for self and the group – were viewed by European colonisers as 'savage'. That many Australians continue to see First Nations people and culture as inferior is a symptom of our ongoing cultural derangement.

The project to emancipate First Nations people is not just vital and overdue; it is linked to our own emancipation. When we tell the truth about our violent past and present – when we truly affirm that black lives matter – we are led closer to the ultimate reconciliation: one in which we come to terms with Indigenous brilliance, and accept that in many ways, the cultural practices of First Nations people are superior to our own.

The Black Lives Matter movement in Australia offers all of us a second chance. Even the devastation of the fires is an opportunity to learn. Here we have the oldest continuous culture on the planet; people who have developed sustainable practices over countless millennia. 'These catastrophic fires that have just happened has woken this country up,' says Indigenous fire practitioner Victor Steffensen, who is now teaching fire services how to properly burn country to protect it. 'Start looking after the land. Look after your rivers, your water. Burn your country the right way. If we see this through the Indigenous lens, then climate change is an exciting time, an opportunity.'

The original Greek word for apocalypse – *apokalypsis* – does not mean 'end times'. It means 'to unveil'. This is the apocalypse we are living through: a process of unveiling and revealing. Patriarchy is not inevitable. It is not sustainable. If we are to survive and thrive as a species, we must first reveal it, and then undo it: in our systems, and in ourselves.

Jess Hill is an investigative journalist who has been writing about domestic abuse since 2014. Prior to this, she was a producer for ABC Radio, a Middle East correspondent for the *Global Mail*, and an investigative journalist for *Background Briefing*. In 2020, she won the Stella Prize for her non-fiction work *See What You Made Me Do: Power, control and domestic abuse*.

One voice

KIM SCOTT

Fire, flood and plague. Toilet rolls and mental health. Injections of bleach and sunlight.

For a studied introvert with a writer's routines, the duty of social isolation is reassuring. In such a tumultuous year I realise I'm extremely lucky to be a homeowner, securely employed and gifted with – as well as the baby-boomer years – certain contraband privileges of whiteness.

With all that luck, it may seem hard-hearted and selfish of me to insist on telling you what I've missed out on so far in 2020. Not just the *fear* of missing out, you understand, but the actuality. What might have been . . .

In late February I had planned to be part of a small group of people visiting a remote beach on the south coast of Western Australia to continue reuniting an Aboriginal (Noongar) creation story and song with its landscape. Our group would be comprised of Traditional Owners – people descended from those that first created human society in this part of the world – and staff from two conservation groups, Gondwana Link and South Coast Natural Resource Management.

The story had been transcribed by a linguist who'd passed through nearly a hundred years ago.[1] The son of the linguist's informant son would be with us. One of the Stolen Generations, he hadn't heard the story, but he knew of the song that accompanied it: a pack of hunting dogs bursts from the trees. They leap through flames and roll into the sea. Swimming with burnt and stumpy limbs, they raise their heads and bark. See how it goes? The dogs weren't destroyed, they were transformed.

There was fire in the story and song we carried, and at the same time, fire was raging all around the nation and a ferocious local heatwave banned all vehicle movement in paddocks and bush. So that trip was cancelled. We made new plans, vowing that nothing less than a pandemic would stop us. It did.

A few of us had visited the site before, relying on hints from the linguist and his informant along with a few scattered memories of where the story belonged. I remember a little over half a dozen of us walking to the far corner of the beach's sandy smile. Sheltered just beyond the calm sea's reach, a fishing net had rolled itself up into a rocky crevice and filled with sand. Some birds had made a nest. 'Oystercatcher' is their common name, 'Kooran-kooran' in Noongar language; you often see them quick-stepping along the ocean's edge.

I hadn't realised it was a nest, but as we got closer, we saw the young birds staring back at us even as we crouched and spoke to them. Someone reached out a hand. Too young to fly, the birds skipped away along the rock sheet and disappeared between rocks in the shallows. They're very good at hiding. Opportunistic, too; their parents made that curious, hybrid nest near the base of the very slope down which the flaming dogs rolled.

The sandy dunes either side of the headland were crowned with a thin layer of crumbly red rock. It might have been evidence, or memory, of fire. A boulder, half-buried at the edge of those dunes could have been a seal, one of those that didn't make it, arching upwards in its death throes. We scanned the ocean, looking for live seals.

In that old story, in that very landscape, a man lit a fire that began in a circle around his errant dogs on the hilltop, then marched down the slope to the sea. It was a very controlled fire.

Controlled is not a word you'd use to describe Australia's 2020 bushfires, our *wildfires*. They howled and devoured like fiery beasts themselves, they incinerated and razed.

Mallacoota looked like hell: gloomy red-tinted light, broiling clouds of smoke and people scrambling into rescue boats to escape the flames.

It's tempting to think of 2020's floods and fires and pandemic as retribution. Many of the world's creation stories — myths — feature punishment of sinners (blasphemers, heathens, infidels, deviants; all those *others*) so that the 'righteous' and chosen can lead the way.

Perhaps. Or perhaps the 'righteous' might be the very same ones who led us into this mess in the first place.

Then again, Australia — as the late great poet Les Murray informed us — is all drought and fire and flood. 'We'll all be rooned,' wrote another Australian bard.

Complacency and fatalism won't help.

Our crisis is not just because of flood and fire and COVID-19. The quality of our leadership as of late 2019 may help explain it. Don Watson expresses it best, with his straight-talking distrust of 'weasel words'. Our leaders are generally 'duck-shovers and deadbeats, incompetents, and cowboys' who comprise 'a bunyip aristocracy . . . besotted with power, sectional interest and the dogmas of economic rationalism'.[2] In addition – and I go beyond Don Watson's words now – our society scoops profit from quarries and prison-industrial complexes, builds border forces against refugees, is stingy with the homeless and battered and mentally ill, and only begins to value public institutions of health, education – and, indeed, of democracy itself – when our dependence on them becomes obvious.

We have the facts. Despite the quibbling and distractions, we know that climate change contributes to environmental disasters like fire, and diseases such as COVID-19. We also know climate change is caused by the damage we have done to the interlocking, natural systems of the miracle that is our planet.[3]

I don't wish to deny the dangers and suffering of 2020 so far, but I'd like to think that good may yet come. Global greenhouse-gas emissions fell over the time of virus lockdown. Away from the screens and hustle there was often a relative calm, a lack of crowds, a change of pace. Rearrangement of transport and work protocols came under consideration, and cycling became more popular. We reflected on the importance of family and friends, noticed the birds. There are new heroes, and new ideas of what and who is 'essential'. We cheer health workers, rely on teachers and carers and have, as Don Watson writes, 'fallen for the scientists, the knowledge flowing on our screens, the honest reasoned arguments and, no less, the varieties of genuine humanity we have been seeing. The citizenry has sprouted heroes.'[4]

After the fires, but before the pandemic, Thomas Keneally hoped for a change in the national response to climate change. Riffing on the relationship between fire and many Australian plants, he said the fires might 'germinate' action on climate change.[5] A veteran of an earlier deadly virus reminded us that COVID-19 might give us the chance to 'take the best possible advantage of new opportunities while preserving and extending the principles of Australian social democracy that have served us so well.'[6]

That same social democracy has recently turned its back on immigrants, international students, casual workers, universities, tourism, the arts . . .

Unprecedented times?

The fires have gone, but they'll be back. The virus lingers, ready to flare up again. I began thinking about this essay in late May, on Sorry Day in Reconciliation Week. 'In this together' was the theme. Our leaders mouthed similar phrases.

An Aboriginal site in the Pilbara was blasted into nothingness. A single plait of hair from 4000 years ago, woven from strands of hair from several different ancestors, was as good as unravelled and cast to the wind. One expert compared it to the destruction of sacred artefacts by Islamic State.[7]

A man in the USA repeated the exact words – 'I can't breathe' – earlier used by David Dungay Jr in Australia. Both were killed by law-enforcement officers. Both knew racism and imprisonment. The proportional incarceration of Aboriginal people is even higher than African Americans in the USA.

A twentieth century poem by Jack Davis about a prison death has the refrain, 'A concrete floor, a cell door, and John Pat.'

Since the Royal Commission into Aboriginal Deaths in Custody thirty years ago there are now more than 430 more Aboriginal names that could be inserted at that end-of-the-line.

Racism burns like a pox and a plague, and is incubated at the centre of how we live and organise ourselves.

We're not all in it together.

It's probably not a matter of a tweak or mere adjustment. A vaccine *might* do it. Reconstructive surgery? Can such a thing ever be minor?

I began with an Aboriginal story of fire. The award-winning book *The Biggest Estate on Earth* tells us that in Aboriginal stories, 'the tone is theological, the teaching ecological':

> A mobile people organised a continent with . . . precision . . . They sanctioned key principles: think long term; leave the world as it is; think globally, act locally; ally with fire; control population. They were active, not passive, striving for balance and continuity to make all life abundant, convenient and predictable.[8]

Fire was used to shape the landscape in order to 'channel', 'lure' and 'persuade' resources to gather at places of easy access. A 'rich spiritual life' and 'time spent nourishing the mind' produced a 'voluminous and intricate' art.

Aboriginal heritage and the natural environment need to be at the centre of any national reconstruction. In *A Rightful Place: Race, recognition and a more complete Commonwealth*, Noel Pearson says there are three parts to our 'national story': Aboriginal cultural heritage, the inheritance of British institutions, and our multicultural achievement.[9]

Put it like this: Aboriginal cultural heritage is a major denomination in this country's currency of identity and belonging. It includes the experience of racism. For most of our shared history, its classical form has been denigrated and almost destroyed, along with its connection to a descendant community. It needs investment and fair rates of exchange. At the same time, I have some respect for the Western canon and Sir Francis Bacon's words that any currency, like shit, is 'of very little use except it be spread'.

And my little group from the beginning of this essay? Aboriginal-led, we want to regroup and share our blossoming heritage with the region's wider community, who, to judge by their 'Kukenarup Memorial'[10] and efforts at land rehabilitation, are trying to come to terms with their own part in our shared history. It might be a model of reconciliation for the nation. However, the denigration and attempted destruction of Aboriginal heritage and the trauma, turmoil and disempowerment of its community are connected. While there's potential for healing and a transformed relationship in processes of consolidating and sharing a pre-colonial heritage, the neo-liberal context of dog-eat-dog and the primacy of the market means there's also plenty of opportunity for a community to be split and to destroy itself over brokerage rights. So here's another fact: it's hard to reconcile and restructure without respectful, supportive infrastructure.

We need the Uluru Statement, and a Voice.

Kim Scott is a proud Noongar man and two-time Miles Franklin Award-winner. He is a Professor of Writing in the School of Media, Culture and Creative Arts at Curtin University.

About the birds this spring

REBECCA GIGGS

London: Mid-winter, and I dream of black cockatoos. Dreams lacking vision, sonic dreams of startling intensity. At dawn the rusted-hinge sound of the cockatoos swerves away, carried off by whatever force in the unconscious revokes the symbols of dreams by daylight. I have been based out of the UK for several months. Having passed through a blue solstice into January, the news bulletins from home now chronicle unprecedented fire fronts, razing tracts of the eastern seaboard to embers and ash. Columns of smoke expunge the stars. I see that in only the barest elements of topography and geology will the landscapes I return to resemble the places I left.

Inch by inch, the immensity of the incineration is disclosed by pull-refreshing my feeds, the daily liveblogs and apps. Red square, sky. Yellow square, inferno. That a disaster staged over thirty million hectares[1] should be delivered in the palm-sized dimensions of online media seems both to spotlight the enfeebling structures of contemporary technology, and to recall the composition of bygone catastrophes, minced through more antiquated modes of communication. I think of battle fatalities relayed by fax; or reports of a hurricane's deadly landfall, fragmented into telegrams. But maybe this is how all truly modern disasters appear, at first: the scale of the destruction proves so singular it exceeds the architecture required to describe it.

Later, I will recognise the obvious subtext of the darkened dreams, that nightly soundtrack of cockatoos. Where the peeping of songbirds ushers in the changing seasons across much of Europe, conventional wisdom holds that the squalls of black cockatoos in Australia announce a more immediate shift in the weather – the parrots cry out before a rainstorm. If there is truth in this claim, it may be because the cockatoos, being sensitive to atmospheric fronts, call to gather together and work at tree bark softened by the rain. Some species eat grubs, which they dislodge from the sapwood using their beaks like secateurs. But it might also be the case that the cockatoos' voices only *sound* louder in air of rising humidity, fenced by low, dense rainclouds – the surrounding conditions, then, not the birds' calls, herald the advancing meteorology.

Either way, as summer glared on in the southern hemisphere, who wasn't desperately pinning their hopes on a downpour? As the months went by, glossy black cockatoos were beginning to surface in Melbourne,[2] where they hadn't been seen for more than 150 years. The cockatoos had flown further than 400 kilometres from the fire-ravaged town of Mallacoota, in search of she-oaks and casuarinas – trees with seeds the birds eat, turning the pods with their fiddling tongues. They may have proved a marvel to those who glimpsed them in the outer suburbs – the orange flash of the plumage beneath their tails passing overhead – but the cockatoos' arrival signalled tribulation elsewhere. So many blackened birds, set alight or suffocated on the wing, also washed up on the coastline along East Gippsland, smudges of colour in scorched debris.[3] Seagulls feasted on dead honeyeaters while the burnt-out bushland fell silent.

As the vegetation comes back, its sonics will be different. The voices of magpies repeat the sirens of fire trucks,[4] while lyrebirds, proximate to the destruction, have been overheard emulating the engine-chuff of heavy vehicles, rolled in to clear smouldering timber. Raptors have done well off the carrion of burnt animals and the visibility created by a reduction in canopy cover, though their populations will begin to starve (or relocate) because of the decline of small ground-dwelling mammals. In time, the first birds to thrive in the regrowth will be hardy and versatile – likely those omnivorous species capable of surviving off a range of immature plants and varied buglife. Endemic birds that feed on seeds or nectar from mature trees, of limited range, will fare badly; as will those that rely on specific kinds of larvae, caterpillars and worms, since fried in the topsoil. More broadly, the loss of insect biomass will have knock-on effects on pollination, and the disappearance of herbivorous animals can tilt the balance towards plants that were previously checked by their grazing appetites. Invasive, fast-germinating weeds[5] inundate the cradles of slow-to-bud 'reseeders' like banksias. Today, volunteers and conservationists uproot the leggy invaders; gardening the wild in a piecemeal fashion. But inevitably the flora that shoots up will be different in composition, creating dissimilar niches for birds. So even after the greenery returns, this new bush will not sound like the old one.

In the early weeks of the COVID-19 lockdowns, people in London reported noticing more birdsong. Stood in the cool of walled gardens and up on tower-block balconies, or during daily, state-sanctioned jogs, Londoners found the dawn chorus unusually – remarkably – loud. 'An *orchestra* of *therapy*,' the naturalist Chris Packham declared on the BBC's *Springwatch*, his palms turned upwards in a saintly gesture to the ruffling green canopy above.[6] Not only in the English capital, but all around the United Kingdom a dormant curiosity for birds was unfolding. BirdWatch Ireland announced its web traffic was up 350 per cent.[7] Online sales of wild-bird food and little suet balls spiked. Near where I was staying, wire bird-feeders appeared like devotions from a resurrected festival; you saw canisters of seed strung up on rooftop washing lines, along the eaves of locked shops, and in alcoves beneath overpasses. Bookstores ran low on field guides.

Whether there were, in fact, more birds to hear was an open question. It was logical to suppose that a boom in bird numbers might track the suppression of commerce, but the precise workings of that consequence were uncertain. Were these countryside birds, emboldened by the motionless streets to join their urban brethren, plucking at lawns and shillyshallying in the ivy? Were more eggs laid and fledglings brooded in borough treetops, now that the smog had waned and the roaring roadways were lulled? With no evidence to corroborate these speculations, their appeal nonetheless summonsed the pandemic's nascent zeitgeist. Each day that I queued for groceries – in Tower Hamlets, a neighbourhood with one of the highest death rates in the country – my eye snagged on a child's careful, open-handed printing. MOTHER EARTH DRAWS A BREATH: the caption on a drawing, taped to the store's frontage. Around a crayon planet, little cars drifted off into outer space. The bird thing, it was a recovery fantasy. One among many that the city nurtured as the shutters came down and the face masks went on. If death meant to take a seat at the table every night, frightening the children, then what else did we have to offer but a fable of nature's resurgence? Beyond the window, see there, hopping about: *so many birds.*

If you had been, before now, a person who identified as a twitcher, you might have observed that these springtime months are always marked by an influx of migratory species flying into Britain from sub-Saharan Africa and Southern Europe. Turtle-doves, pipits and nightjars, martins and warblers come wheeling in on the ocean breeze. The British public pin the calendar on hearing the first *moot-moot* of a cuckoo,[8] and people celebrate the day, months later, when the earliest swifts are sighted;[9] a burst of arrows, dropping through the clouds. By mid-May, a new grammar of movement animates the roadside. Hedges pop with flycatchers and finches, their tiny hearts zinging left and right like mustard seeds flung into hot oil. On the canals, visiting waterfowl unzip the reflected sky, while swallows overhead settle the tally of their ascension on the powerlines. Blank space everywhere – the open air, a lake's glassy surface – gets hatched and crosshatched, iteratively, by feathered life. These thousands of journeying birds, pouring in from thousands of miles away, arrive full-throated with a repertoire of courtship songs; it is their breeding season. Springtime is bookended by the appearance of new species and on this account, it seems, the English year surges forwards by force of avian momentum – as though the birds didn't follow the weather, but delivered it. And perhaps this was a part of it too; the burgeoning enthusiasm for birds and birdsong in 2020. Now that the days themselves felt shapeless, the future stalled or upended, people hungered for visible propulsions of time in nature, to which the comings and goings of birdlife could be recruited.

So, yes: there were more birds to hear in London during lockdown, but no greater flourishing of their populations than usual. It was the context that had changed. Interviewed by reporters, acousticians pointed out that the birdsong might only sound louder this year, because the background grind of machines had dropped away. 'You can hear into the distance,' said wildlife recorder Chris Watson.[10] Birds-near could be made out in concert with birds–afar; distant birds, hitherto occluded by the noise of motorways and flightpaths. For this reason too, the ear was fooled into perceiving the birds as more numerous, and nature as uncommonly profuse.

To say that the setting in which birdsong was heard was novel is to acknowledge a state of affairs beyond the hushed ambience of the lockdowns. What people wanted birdsong *to mean* had also changed: birds were burdened with fresh significance. If the viral threat came from nature (SARS-CoV-2 being a 'natural' organism), then the fact that nature also still possessed idyllic, sing-song dimensions proved soothing. The pandemic might ravage the hospitals, but there would remain woodlands, berries agleam in the hedges, and things to flutter and soar. Bridled by the panic of midafternoon you could send your mind out, like a balloon on a string, to float with the collared doves. I, for one, am not immune to the localised miracle of a nest with eggs. A nest, as I see it, is not just an environment inhabited by another form of life – it is the bird's own sentience made material: its brain processes, looped upon a branch. Like many people, what I instinctively wanted from birds, in this moment, was proof of some other, overlooked world nestled into the human one; a place from which all our troubles might then be viewed, through the bird's-eye perspective.

If what people heard, when they meditated on birdsong, was evolving that spring – so too was how they were listening.[11] In contrast to emergencies borne of military conflict or extreme weather, the COVID-19 pandemic seemed notable in how it marshalled people's hearing, above any other faculty or sense. Midmost in this crisis were the sick, who typically marked their illness's onset by a loss of the ability to taste and smell (giving the impression SARS-CoV-2 worked like a jewel thief: disarming perimeter sensors first before moving on to the plush red cushioning of the lungs). But in surrounding communities of the well, a new and pervasive pathology of sound had also arisen.

Nurse practitioners, manning hotlines, concentrated for telltale hoarseness and consistency to diagnose the dry COVID-19 cough. As the rumblings of industry and transport ebbed, ambulance sirens and the *pit-a-pat* of medical helicopters stood out like aural beacons, disclosing a hidden landscape of suffering. During plague and epidemics of the past, churches of many denominations stilled their bells on the grounds that the peals (some were death-knells) tended to agitate the infirm and dying.[12] Now, the belfries fell silent because bell-ringers were wary of congregating. Only the *adhan* continued. On loudspeaker from an east London minaret, brocading the empty air, the majestic *adhan*; hastening those observing Ramadan to prayer inside their houses, and filling every corner of my own godless heart. Some nights the windows were thrown open and neighbour joined neighbour to applaud the National Health Service. Other nights, neighbour heard neighbour in bitter argument through the wall. Overwhelmingly, though, there was just this: *the quiet*. Hours of quiet. A thicker kind than before. On a hot evening you might have believed it possible to piece out a raven's wingbeat into slow, separate brushstrokes: to hear flight-feather, then glide-feather, and down. In all this stillness we were straining our ears to apprehend how bad things would become; to hear in what direction, and how fleetly, the crisis gained upon us.

Plagues are like this. They derange the senses[13] in a symptomology that gallops ahead of the infection. Around this time, news of a strange figure began to surface in north London and Norfolk – a person with the head of a crow. In Crouch End,[14] and also in the Norwich suburb of Hellesdon,[15] watchful locals reported spotting individuals promenading in outfits distinctive to seventeenth century outbreaks of bubonic plague (the Black Death).[16] The eccentric costume consisted of a long-beaked mask – like a dead bird's sun-stripped skull – beneath a wide-brimmed hat over a floor-length overcoat. The masks recalled humanoid fabulations in Hieronymus Bosch's *Garden of Earthly Delights* triptych, or perhaps Thoth, the ibis-headed Ancient Egyptian god who commanded the arts of writing and science, and stood in judgement of the recently perished. In the highly codified cast of the Italian *commedia dell'arte* there, again, is the bird-headed figure. The curved beak has something of the Grim Reaper's scythe. Befittingly, at Venetian masquerade balls the character was known as a memento mori – death upright among life. But the ensemble has, in fact, more pragmatic, medical origins.

During the Great Plague of 1665–66 many Europeans fallaciously believed disease was transmitted via 'miasma' – foul odours released by rotting organic matter and sewage. So doctors protected themselves by wearing long nose-cones stuffed with pungent herbs including lavender, and camphor, and sometimes enclosing whole dried roses on the stem. The bird-faced mask is PPE from a time before the word 'virus' existed. Rather than being 'read' from outside as a visitation from some underworldly being, the mask's function is internal, being shaped to concentrate the mask-wearer's olfactory sense and deodorise pestilent air. With hearing muffled (a head wrapping was worn beneath the mask) and patients viewed through fogged lenses, smell was the primary faculty through which that pandemic was modulated – for doctors, *and* for those who feared the infections of miasma as well, who might also carry sponges soaked in vinegar to whiff, or place fragrant oranges pierced with cloves around the home.

To the plague patient of history, the bird-headed person wore the livery of a healer. Relief, not fright, then, met them at the door. But seen afoot in 2020 these plague-doctor capers scanned as portentous: people called the police (who, for their part, offered 'words of advice' to the pranksters). The garments implied not just that we might fall back through time into the baroque medical theatres of the past, but that adaptations made on account of what we didn't yet know about COVID-19 would come to look wholly bizarre in the future. The fear, I thought, was that today's science might yet come to be viewed as superstition.

True, the logic of superstition can furnish a false fantasy of recovery, or one of escape. But in the midst of the lockdowns, the sounds and sights of birds reminded me, most of all, of the extent of our connections to one another. For the plague doctor's mask is evidence, to my mind, not of division from the sick, but of ministration and aid – while the calls of cockatoos threaded news of firestorms through Melbourne suburbs, and my own sleeping dreams.

During those months I often spent evenings seated on a square of astroturf someone had laid out on the bituminised rooftop above the flat I was living in. As the sun descended, I watched flocks of birds dot and dash the horizon, manoeuvring within rills of weather invisible to me. After a certain point the birds dissolved into twilight, and then, very gradually, I would begin to notice the lights of living rooms and kitchens below. In the darkness I saw how many of my neighbours stood at their windows, listening, with me, for what moved in the air between us.

Rebecca Giggs is an award-winning writer from Perth, Australia. Her work focuses on how people feel toward animals in a time of ecological crisis and technological change. Her debut book *Fathoms: The world in the whale* was published in 2020.

The coming storm

RICHARD McGREGOR

Forced to move house during the lockdown – legally, I might add, as lifting boxes was mercifully classified as an essential service – I stumbled across an old poster that seemed eerily in tune with the times. The poster had been plastered around inner-city Melbourne in late 1978 in the days after Sir Robert Menzies died. PIG IRON BOB DEAD AT LAST the headline blared. A smaller strap across the top read: OBSCENE IMPERIALIST RITES FOR MILITARIST WITCH HUNTER. A bushy-browed Menzies was shown waving his right arm, an image that had been cropped to give it a whiff of a Hitlerian salute.

Looking afresh at the poster after so many years, my first instinct was to admire the agility of its North Korean-style invective, vituperative insults and historical fury all jammed into a few words. It was only a few days later, when one of Menzies's Liberal Party successors, Scott Morrison, got to his feet at the Australian Defence Force Academy in Canberra for the country's annual defence strategic update that the full extent of its contemporary resonance hit me.

Speeches like the one delivered by Morrison are designed to peer over the horizon to threats that look uncertain and unformed today but could materialise into something serious in decades to come. Morrison was anything but vague in his depiction of the coming storms in the Indo-Pacific. He went back to the future for guidance, mentioning the 1930s four times. It's an era, he said, that he'd been 'revisiting on a very regular basis, and when you connect both the economic challenges and the global uncertainty, it can be very haunting'.

A cartoonist illustrating a column about the speech drew the prime minister looking in the mirror and seeing Winston Churchill. Much as Morrison might have been thrilled, the comparison struck a lazy note for me, with, in the apparent absence of a local hero, a touch of cultural cringe as well.

Before COVID-19 — before, in other words, the first wave of mysteriously stricken patients began to crowd into hospitals in Wuhan, and the lies and the chaos and the lockdowns that followed as the virus spread out from the giant inland city to the rest of China and then overseas — I used to make lists of all the things that Australia and China were battling over. Cyber attacks, Taiwan, the South China Sea, Huawei, the Pacific Islands, Hong Kong, foreign interference, universities, Xinjiang, Australian prisoners held in China, Crown casinos, raids on journalists in both countries by secret police. As soon as one issue drops off the front pages another one rushes to take its place. Even the two countries' sports stars were fighting, with champion swimmer Mack Horton accusing his rival, Sun Yang, of being a drug cheat.

The virus lifted the mutual acrimony to a new level. In Canberra's telling, the Australian government's call for an independent investigation into COVID-19 was an entirely reasonable response to a pandemic that had cratered the global economy. In Beijing, the Foreign Ministry denounced it as a 'political manipulation' made at the behest of Washington. Or, as one Chinese netizen put it, Australia was 'this giant kangaroo serving as a dog of the U.S.' Beijing announced trade sanctions on beef and barley to drive home its displeasure. Both nations warned their citizens that travelling to each other's countries was dangerous. It didn't really matter, as no one could travel anyway, but in terms of political signalling, it was potent. This was a relationship heading downhill fast.

To get a sense of just how bad things might get, let's go back to Australia's longest serving prime minister, and the disrespectful send-off that some old-fashioned socialists in Melbourne provided for him in 1978.

Menzies was given the nickname 'Pig Iron Bob' by waterside workers in 1938. They were protesting against the sale of pig iron to Japanese steelmakers, which Menzies, as the then Attorney-General, forced through in the face of industrial action. Coming just after the Imperial Army's massacre of Chinese in Nanjing, the waterside workers objected to Australian resources being used to feed the Japanese war machine. In that respect, the 'Pig Iron Bob' posters didn't make sense. Far from being a 'militarist', the criticism of Menzies was that he hadn't, at that point, been hawkish enough. Nonetheless, once the Pacific War was underway and Australian troops were on the frontline battling the Japanese, many conservatives, with the benefit of hindsight, came to sympathise with the communist union. Menzies himself, on the left at least, struggled to live the nickname down.

These days, the Australian government has a close relationship with Japan and views China as the major threat to peace and stability in the region. This, by the way, drives Chinese diplomats in Australia around the bend. It doesn't take long once you meet for them to remind you that China and Australia were allies in the Pacific War. That's certainly true, though the diplomats don't like to be reminded in turn that Australia was allied at the time not with Mao Zedong's Communists, but Chiang Kai-shek's Nationalists, who, after losing the civil war, fled to Taiwan, an island that remains unfinished business to this day. The diplomats also seem blind to the fact that China under Xi Jinping has much in common these days with the Japan of the 1930s.

But back to pig iron, which, like iron ore, is an essential ingredient in making steel. Luckily for Australian resource companies, and the Treasury, which collects taxes from them, China's economy is still heavily reliant on building things which use steel. Even though its share of global economic output is about 20 per cent, China makes more than half of the world's steel. As a journalist based in Tokyo in the 1990s, I had to track the annual iron ore price negotiations between Australian miners and Japanese steel mills, at that time by far our biggest customer. Analysts in Tokyo and around the world would pore over the latest signs from local steel mills of an uptick in production, which at that time averaged in total about 100 million tonnes a year. Anything more would be a bonanza for the miners. China, however, has reached into another dimension. It is making nearly one billion tonnes of steel a year, to build new cities and infrastructure, ten times what Japan used to produce.

For a country like Australia, located nearby and sitting on huge reserves of high-quality resources, China's steel boom has been a once-in-a-lifetime windfall. China is Australia's biggest export market, and iron ore sales comprise the biggest component of that. Even as the pandemic persisted in much of the world, Australian iron ore exports were reaching record levels in 2020 as China's economy recovered and steel production along with it. The shiploads leaving Western Australia for ports along China's east coast have been fetching fat prices too, as production from mines in Brazil, Australia's main competitor, were disrupted by accidents and the coronavirus.

So, when Morrison looks in the proverbial cartoonist's mirror, does he worry that one day, instead of seeing a wartime hero like Churchill looking back at him, it will be a pre-war Menzies in his place? In other words, does he worry that China's relations with the West will become so bad that his job as prime minister will be not to reap the benefits of iron ore sales, but instead face the prospect that he might have to restrict them? 'Steel Scott', anyone? I know it doesn't have quite the same stench as 'Pig Iron Bob' but it could take on a more sinister ring in the hands of a skilled political opponent.

Morrison has hardly discouraged such hyperbole. After his speech at the defence academy, he told the *Sydney Morning Herald* that he had deliberately thrown in those multiple references to the 1930s because he thought the Australian people weren't sufficiently aware of the potential dangers ahead. 'That's why I thought it was important to stress the point,' he said. Morrison wasn't just thinking of the end of that decade, but its opening years as well, which ushered in the Great Depression and a sustained economic downturn, which in turn contributed to the conditions which created the world war that followed.

Before COVID-19, such grave historical references wouldn't have passed the laugh test. Slow the sales of our single most valuable export, at a time when we need economic growth and tax revenues more than ever? But don't think that the idea of Canberra interrupting iron ore sales to China hasn't crossed the minds of senior executives in the C-suites of our big miners. Nor have the Chinese missed the signals themselves, which is why they are focused on securing alternative suppliers for core industrial resources, by reviving iron ore mines in Guinea, West Africa, as well as cultivating closer ties with Brazil.

Naturally, the high-stakes debate over China lends itself to extremes. On either side of this debate, accusations that single out people as either warmongers or traitors are commonplace these days. The debate has been transformative in other ways, too. After decades in which the idea of strong government has been under sustained ideological attack in many western countries, state power and state capacity are back in vogue.

China, of course, never bought into the small government movement and has always been committed to building and maintaining a powerful state. After mishandling the start of COVID-19 in Wuhan, the Communist Party in China clicked into gear at a frightening tempo to bring the virus under control. Without the need for any messy democratic debate about civil rights or so forth, the government was able, almost overnight, to lock down more than 700 million people in residential detention; seal provincial, city, county and village borders; shut factories while commandeering the entire output of some businesses to supply emergency medical equipment; order the wearing of masks; mobilise military and paramilitary units; build pop-up hospitals; mandate testing of tens of millions of citizens; and track the movements of residents using mobile phone apps. The mobilisation of the state, businesses and people at such short notice was a potent reminder that the ruling party effectively has war powers at its fingertips in any declared emergency, even in the absence of conflict with a foreign power.

The conservative Morrison government has had to embrace the state in Australia during COVID-19 and try to plug gaps in strategic industries, with little thought for budget deficits, all very much the hallmarks of a wartime administration. Australia also created a national cabinet to manage the crisis. China, of course, as a single-party state, has had one all the time.

For some, the idea of perpetual conflict with China, overlaid by the miseries of COVID-19, is an exhausting prospect. For others, the prospect of standing up to China looms as an exhilarating, clarifying experience. The response of some Australian politicians to the China challenge, and their enthusiastic, unyielding belief that Australia has reached an historic turning point, reminds me of the way that friends of the late Christopher Hitchens tried to explain the hitherto left-wing polemicist's fervent support for George W. Bush's second Iraq war. 'He's always looking for the defining moment, as it were, our Spanish Civil War, where you put yourself on the right side, and stand up to the enemy,' one of Hitchens's friends, the writer Ian Buruma, told the *New Yorker*. Hitchens himself foresaw 'a war to the finish between everything I love and everything I hate', and a question on which history would judge him.

Andrew Hastie, the Liberal backbencher and chair of the intelligence committee, is one case in point. He arrived in Canberra with the views that you might expect of a veteran of the Special Forces' mission in Afghanistan: believing the so-called war on terror spawned by the September 11 attacks to be the paramount challenge of the twenty-first century. But learning about China has been a game changer for Hastie. He has come to see Xi Jinping as a latter-day Joseph Stalin, and Australia as complacent as France before the Nazi invasion – and that was before COVID-19.

The likes of Morrison – and Hastie – have made up their minds about which direction the world is heading. Their views aren't perfectly aligned, and while Morrison occupies a position of genuine power, Hastie is a backbencher. But in different ways, like many of their colleagues, they see themselves as the guardians of their country and its values at a hinge point in history. Hopefully, their strategy of standing tough alongside allies against China, which is one that you see emerging around the developed world, will deter the rising superpower.

Otherwise, we can only hope they are wrong about China. It is all very well to want to look in the mirror and see Winston Churchill staring back at you. Churchill was certainly a hero to many, but only after one of the most destructive wars in history.

Richard McGregor

Richard McGregor is a senior fellow at the Lowy Institute in Sydney. He reported for decades in Japan and China and is the author of numerous books on Chinese politics and foreign policy. His book on Sino-Japanese relations, *Asia's Reckoning*, won the Prime Minister's award for best non-fiction book in 2019. His last book, *Xi Jinping: The Backlash*, was published in 2019.

A convincing darkness

OMAR SAKR

In the pre-dawn hour, as the dark becomes less sure of itself, I get up to go to the hospital. It's cold, and I have to fast before this surgery, so I take my Lexapro pill with a sip of water right away. I had been diagnosed with cholesteatoma before the pandemic locked the world in its grip, and gone onto the public waiting list. The ENT specialist told me it was a routine surgery, a cutting away of abnormal growth behind the ear canal; decades ago, people died from this, their own skin growing into their brains. There was some risk of deafness, or damage to a nerve that could paralyse half of my face, but he had never slipped yet. I'm the kind of man that assumes such odds exist to spite me, so I was not reassured. My fiancée Hannah drove us to St Vincent's at 6 am, and the roads were busy, maybe because the restrictions were set to ease the next day and people couldn't wait, or maybe because capitalism is a death cult that will brook no surcease, people gotta eat or work to pay the landlords, and we passed the time by shaking our heads at everyone's foolishness as a way of ignoring our own.

The hospital itself was a confusing place. If ever I was to get the disease, it would be here, I thought. Signs sent us to level four, the reception where screenings were to take place for all visitors, except that the entire floor was empty. It was not yet 6.30 am, but the unlit carpeted halls, with large COVID-19 warnings plastered everywhere, and staggered barriers for queuing, was nonetheless eerie. It turns out we'd come up into the wrong building, which was closed, and had to hurry over to the public hospital entrance. A woman sitting at a table in the foyer entrance asked if we had any flu-like symptoms. No? She waved us through with a little flick of her wrist. It seemed woefully inadequate – I'd seen more rigorous screening at the Apple store in Broadway. Later, Hannah would tell me there had been equipment that took our temperatures as we walked by, that security had been monitoring it, and perhaps I missed it.

Part of me doubts that any of it matters, and I had said as much the other day as we paced around the block to get our daily exercise. The restrictions, or processes put in place (to call them restrictions is to invite resentment), were as much symbolic as practical; it was largely performance because, even now, as we did our best to follow them, we could be breathing in or walking through the infected droplets that someone else, equally doing their best to physically distance themselves, had left behind on their walk. We want to feel like we can control our own destinies, and so we pretend that what we do ultimately matters. As a Muslim raised by high school dropouts in Western Sydney, I often heard this particular kind of fatalism:

'If Allah wants you to go, you're gone, there's nothing you can do about it,' to which I would say, 'Okay, so go play in traffic right now. If it's not your day, it's not your day,' and yet now, I found myself on the other side of it, giving in to the awesome omniscience of God. It was easier that way, and absolved me of any responsibility. The pandemic, invisibly everywhere, had soaked such hopelessness into my bones. Which is why I'd been affronted by the screening. The acting was subpar, the role of the process almost subliminal in an age of intrusive security measures that were as equally and vehemently based on my body. I wanted to be comforted, and received nothing.

Eventually, we found the right admission area. I filled out some forms, answered questions in triplicate posed by nervous students and a tired nurse. I was measured and weighed. I am 6ft 1, over 100 kilos. I've been a big guy for years but I had put on weight since my last weighing only months back, and it seemed silly to care about that, but I did. To be your heaviest at death's door makes a kind of poetic sense, but vanity gives zero fucks about poetry – and yes, I was thinking about this relatively routine surgery as a death, because I am Arab and given to full-on dramatics at the slightest opportunity, but also because to go under full general anaesthetic is to register a profound rupture in experience. It is not the same as sleep. Even sleeping, the body knows enough to dream, it shifts and moves, it is a restless elsewhere. There is awareness, however quietened, estranged.

Once you are under, there is nothing. Your ribbon of memory vanishes. You wake in a different place; you can stitch the two moments, pre-injection and post-op, with the word 'surgery' but the word will sink, and only blackness will remain. I was coaxed into consciousness by a nurse. She said the operation went for four hours. I faded in and out of my fuzzy body. I was given a large plastic jug to piss in, and I let out the pent-up urgency. The body, then, had its own recording of life I was not privy to, and I was reassured and disturbed by that, the knowledge that whatever composes my 'I' is immaterial.

I was wheeled into a recovery ward, my head swathed in bandages. I took a selfie as soon as I was physically capable of it, and was suitably impressed with my own damage. A Palestinian friend texted to tell me I looked like a Beirut war victim. Beirut, which has not been in a war for many years — except with itself — is forever associated with it, and I was reminded my body, too, carried this catalogue of bloody images. I watched through a haze as social media reacted to the photo, wondering if the closer you are to the catalogue of images society pins to your body, the more comfort it inspires, the more relatable you become. Cynicism suggests that no one likes me more than when I am hurting in a recognisable, tangible way, but I have to leave room, as well, for sincerity and compassion. I watched as an old man had to be ushered into his nearby bed by a trio of nurses, and I listened as they pleaded for his cooperation at every meal time, for every test, pill or injection; I was awed over and over by their resolute care. This is the best of us, I thought, and felt such a stinging, familiar shame that we do not provide it for all — and worse, that we actively harm the disadvantaged, and dispossessed. What could we become if we insisted on this reverence for life, always, and were not partitioned by the politics of birth, of borders and class?

Hannah interrupted my reverie with her arrival. She was jolted by the turban of bandages, but tried to cover it. 'Your head is hectic,' she said. Only months ago, we'd gone to Inverell to help her father, who is recovering from multiple cancers and still on chemo, take care of some long languishing yard work and physical labour. We had spent our fair share of time last year in hospitals, by his side, and now here she was by mine, the strength and youth of yesterday erased. She helped me cut up my food and did not show what it must have cost her to be so present, and caring, and to have to go home alone at the end. The man opposite me looked to be around my age, a young wog with a leg injury. He had a steady stream of lads visiting him, mostly wogs and Lebs in trackies and Nike TNs, and through the general lack of privacy, the fact that we were all ill together here, I learned that a building site had collapsed on top of him and some other tradies who were in different rooms, so their friends were doing the rounds, generally being rowdy, shooting the shit, trying to make them laugh, every inch of them alive and resistant to the hush of the sick. Their irreverence made me feel at home, and lonely all at once for a life I had largely removed myself from in pursuit of an art form few people give a shit about.

I slept, I ate, I pissed in the familiar plastic jug, having lost virtually all control over the condition of my life. It was freeing in one way, and of course stifling in another: a microcosm of the broader situation in society. My bed was by the window at least, a slice of sky and city mine to behold at all hours. My aunt sent me a text asking if I would get prescribed Endone, and if she could have some, she was in such pain, her own knee surgery had been postponed. My Turkish uncle called to say a relative I never knew had died, he was in his fifties, and sent along a photo of the man with my father, who was also dead. They were waiting now for the death certificate, and for word on how many could attend the funeral. I took in what I could, I let go of what I couldn't, and all the while the nurses, mostly brown Southeast Asian men and women, tended to the bodies with their immaterial, essential hosts. I want to make of them and this work a sacredness, even knowing there is little romance in the gruelling hours, dealing with the stink and rot of failing flesh. I am grateful for it, regardless.

I've been home for several days. No plastic jugs anymore, thank God, but otherwise little has changed. I mostly sleep, and I am lucky, so lucky, to be cared for by one I love, to have work in writing I can continue with from my bed, or the short walk to my desk. I have to take various pills throughout the day, which prevents me from observing Ramadan, the holy month made strange and strained by isolation, and my ear is a clouded knot of pain. There is talk, all the time, of restrictions easing, and not easing, of who deserves to die, of letting the old *go*, of the 'economy' needing to start again, which is to say the rich need to get richer again (a phenomenon that actually never stopped), and a sense already of an acceptable level of sacrifice in order for this to occur. This is evil at its most banal, and it shows no sign of abatement. Let the nurses and doctors suffer, let the labourers build and break, let them all grind their bodies to the mill, for somewhere a bank balance must grow.

The days have become months, and the world has changed far beyond the scope of this essay, with cities, states, and countries opening and closing like anxious flowers attempting to halt the damage of re-occurring coronavirus outbreaks. One thing hit harder than anything else, both figuratively and literally: the 'Beirut Blast', an allegedly accidental explosion of 2750 tonnes of neglected ammonium nitrate that levelled the port, killing more than 190 people, injuring 6500 and leaving hundreds of thousands homeless. It rates as one of the most powerful explosions in history, and it devastated my mother's country, which was already struggling thanks to a corrupt and inept government, with food and power shortages the norm, to say little of the hyper-inflated currency, and a population of more than a million refugees, all amid the pandemic.

Part of me wants to go back and erase the words I wrote the day after I left the hospital, the joking remark about my bandaged head, my wounded state instantly being compared to Beirut, or my comment about it being at war only with itself, but much as they sting, they have proven all too true. How else to explain a government that could leave thousands of tonnes of explosive material in a densely populated area, except as suicide? How else to explain the lack of aid to Lebanon to deal with its problems when its people were starving, and killing themselves in despair, as in the case of a 61-year-old man who shot himself outside a café, carrying the Lebanese flag, a copy of his clean criminal record, and a simple note: 'I am not a heretic.' Years ago, an act like this prompted the so-called Arab Spring. In 2020: some protests from an exhausted local populace, and not much else. Meanwhile, the devastation of the recent warehouse blast has made worldwide news, and led to huge fundraising efforts. Once again it is reinforced that for Arabs to be seen in the West, we must first be linked to a bomb or damage. I am so tired and hurt that I almost don't care about this anymore: see us and hear us however you want, only first, please, help my people survive.

The world has changed, and yet, its worst features persist. We are still being asked to acknowledge this is an extraordinary situation which requires a total change of our behaviours to accommodate it and our survival, but only so long as we are able to change *back*, to a way of life that is not just riven with deep inequalities, but which experts have already determined is fatally flawed for human society. What is the normal to which we are being pushed to return? My normal is the precarious life of a working-class poet in a country that hates him, his culture, his communities. My normal is racist commentary on my work, death threats and hate trolls. My normal is my aunty's broken body, my father-in-law's cancer, my mother's unstable rental situation. My normal is cousins locked in a carceral loop, accustomed only to poverty, punishment and police harass-ment. My normal is life on stolen land, where self-determined outcomes by First Nations communities are ignored and their deaths in custody continue. Death is a passive word here. My normal is the deep privilege of knowing whatever my family or I go through, our kin in Lebanon and Syria have it worse, and we have contributed to that. My normal and your normal is a relentless march to a ruined climate, the dismissal and under-mining of scientists these past few decades, the lack of leadership and vision that dares to imagine a sustainable way forward.

The greed and cruelty, the endless consumption that marks the modern way of life, threatens to overwhelm me constantly, but unlike the deep dark of anaesthesia, this is an unconvincing darkness, and we do not have to stay under it. I admit I have no great hope we will take hold of our destinies and use this chance to transform for the better; I think we'll stay mired in an unnerving mixture of complacency and crisis, but as I mentioned, I take comfort in performance in the absence of control, and writing has always required my best.

Omar Sakr is an Arab-Australian Muslim poet. Winner of the 2020 Woollahra Digital Literary Award for Poetry, his most recent collection is *The Lost Arabs*.

A reinstatement of the facts

LENORE TAYLOR

I have always worked with facts. I have sifted them for relevance, assembled them to make sense of things, and used them to construct an argument or to disagree with another point of view. Facts are, for journalists, the essential ingredient, like flour for bakers or clay for sculptors. So I recall very clearly how disconcerted I felt when I first sensed they were turning to liquid and sliding through my hands.

It was during Tony Abbott's campaign against the Labor government's carbon pricing scheme – the policy he dubbed a 'great big tax on everything'. There were, for sure, some factual arguments that could have been deployed against that policy, or alternative ideas that could have been raised. The then opposition leader opted for neither of these methods. Instead, he travelled the country saying things that were patently nonsensical. But most news outlets reported them uncritically, and this firehose of nonsense proved impossible to mop up.

Most days Abbott would visit a store to make wild claims about price hikes, with only a few journalists, myself included, trailing behind him to unpick them. He went to butcher shops, where worried owners would talk about the looming increase to their power bills.[1] I'd calculate what price rise the butcher would need to pass on that extra cost: less than 0.2 per cent, or 2 cents on a kilo of mince. Since the butcher's low- and middle-income customers were going to receive compensation from the carbon scheme and would presumably have been able to afford this price increase, the impact on his business would be minimal. But by then the nonsense roadshow would have shifted to a supermarket or a pie shop or a pizza shop,[2] where the price of pizza boxes was going to rise by one cent per box.[3]

We all know now that the facts never caught up with that 'axe the tax' campaign, and that some years later, in a truly breathtaking display of cynicism, Abbott's chief of staff, Peta Credlin, blithely conceded that they'd always known the carbon price had never been a 'tax' at all, and they'd only called it that to stir up 'brutal retail politics'.[4]

Despite the inadequate efforts of some of us, the media was a conduit for these 'brutal retail politics' throughout that time. For the most part, Australian journalists continued to play by what many considered to be the 'rules of the game', rules that assumed that politicians would stay within the guardrails of truth save for a bit of spin to spit and polish a campaign, and that, correspondingly, journalists would report what each party said, with any critique coming in the 'reaction' paragraphs at the bottom of the page.

I always believed that interpretation of the rules to be mistaken. That's why I was so disconcerted when properly factual reporting was swamped during the carbon pricing 'debate'. As the US press discovered even more starkly and consequentially during the 2016 presidential election, those rules just don't work if the politicians themselves aren't in any way constrained by facts – if they lie, contradict themselves, or simply make things up, or if their calculated tactic is to upend or subvert civic dialogue rather than participate in it.

From that point, the liquefaction of fact-based reporting gained pace. Not content with abandoning the factual norms of public debate, President Trump set about undermining the very concept, popularising the term 'fake news' to dismiss anything that made him uncomfortable, and using social media to create a closed loop of alternative 'news', based not on facts but on whatever alternative reality the reality TV king had dreamed up that day, and on demonising the 'liberal' media for trying to maintain a fact-grounded public discussion. Facing cost pressures and an audience splintering across the internet's infinite sources of 'news', outlets like Fox in the US or Sky News (after dark) in Australia built a whole business case on amplifying this populist polarisation and disinformation.

Then came the pandemic, where facts were the most important tool to limit the virus's spread, as well as the consequential deaths, economic hardship and horrible human suffering. Here, as around the world, readers responded to the pandemic with a seemingly insatiable need for information – the daily briefings; the case numbers; the hospital capacities; long explanations about everything scientists knew about the virus, everything it did to the body, everything epidemiologists could tell us about how it transmits. They wanted maps and graphs and detailed explainers. They wanted instructions about how to make masks and how to homeschool and how to cope emotionally. They reached for the best available facts and knowledge as something solid to cling to as our former lives and plans dissolved into uncertainty. Like many news sites, our readership soared and stayed there, dipping and rising slightly like a proxy map for the nation's fears and anxieties: sky-high in March, down a little in April, May and June and then rocketing back up again in July as the second wave swept through Victoria.

In Australia, our job was made easier because for the most part leaders followed expert advice and sidelined the shrill voices urging them to do otherwise. With some exceptions and a few wobbles and disagreements, leaders listened to the health experts and imposed lockdowns and restrictions, brushing aside the few commentators suggesting that things be left open in the interests of economic growth, that the elderly would be willing to bear the consequences, that the severity of the whole thing was being exaggerated.

The federal government's fiscal response has pushed the deficit to a size not seen since the Second World War. There is plenty to quibble with in the design and scope of the assistance payments, and even more in the plans for a 'gas-led' economic recovery. But in broad terms, despite the obvious contradictions with its own 'debt 'n' deficit' critique of the Rudd government's response to the global financial crisis, and some nervousness within its own ranks, economic advice has been heeded.

The enormity of this crisis seems to have shocked Australian politicians out of their reflex to turn to the politics of division, and they have instead sought to respond carefully, sensibly and in ways that might unify the nation. By and large, they have worked cooperatively with one another, and managed to accept nuance and maintained the ability to change tack in line with new evidence. The return of reason has, of course, not been universal. There's been bickering and tension, and, as the pandemic dragged on, the predictable attempts at buck-passing blame.

Despite all that, it turns out that old-fashioned evidence-based governing is an effective way to respond to a pandemic.

In the US – where President Trump has ignored warnings, sidelined public health experts, insisted the disease will just disappear 'like a miracle' and spread innumerable falsehoods and conspiracy theories – the mortality rate, as measured as deaths per 100,000 citizens, stood at 51.8 at the time of writing. (Although, in an interview with Axios's Jonathan Swan, the president himself did not seem to understand these figures.) In the UK, where Prime Minister Boris Johnson began by boasting that he had shaken hands with everyone in a COVID-19 ward before changing tack after being hospitalised with the disease himself, the rate was 62.4.

In Australia, where state and federal leaders have cooperated, followed advice, held every press conference alongside public health officials who have themselves now become trusted public figures, and have for the most part followed expert advice, the rate is 2.1, although likely to rise, given the Victorian outbreak. In New Zealand it stood at 0.45.

As well as being a better method of dealing with the most immediate challenge of our lifetimes, self-interest may also motivate politicians to maintain this change, because, it also turns out, voters like it. By August 2020, Scott Morrison's handling of the crisis was approved by more than 60 per cent of Australians, and the approval ratings for state premiers were similar. Victorian premier Daniel Andrews has been under sustained attack from sections of the press for his state's mistakes far more fiercely than any other leader. But despite this, and even as he sought to manage Australia's first stage-four lockdown and Australia's worst transmission rates, his approval ratings remained at around 50 per cent.

And when the political debate stays within the broad parameters of sensible discourse and reasonable disagreement, it is far easier to report and interrogate. We didn't have to explain why the leader of the country was wrong to suggest that the virus could be cured by injecting bleach, or decipher daily contradictions and incoherence.

Sure, Australia has its anti-maskers and 'sovereign citizens' and anti-vaxxers and various other people who see some kind of conspiracy behind the roll-out of 5G, but the overwhelming response from journalists, politicians and public health experts is to condemn them as dangerous cranks. Conservative columnists can exercise their right to fulminate about how compulsory mask-wearing is 'virus hysteria' and others can describe lock-downs as tantamount to 'dictatorship', but the vast majority of Australians are politely declining to engage in this particular culture war in the interests of getting on with following expert advice and fighting the spread of the disease.

Early evidence suggests this has enhanced people's trust in the Australian media,[5] and that, unlike in America, where the idea of news has been politicised, there isn't a huge partisan difference in the extent to which Australians trust the news.[6]

Could it be that expertise is making a comeback? Could it be that we are able to create a virtuous cycle of politicians being rewarded for dialling down hyper-partisanship, and the media rewarded for a somewhat calmer tone to reporting and analysis? Dare we hope that facts might be solidifying again as the sensible, common-ground foundation for civic debate?

There are increasing problems alongside these possibilities: the pre-COVID pressures on the media business model are only accelerating, and the global, mostly unregulated 'open mic' of the internet continues to present news, and credible citizen commentary and views, alongside outright lies and disinformation as if they are equally credible.

To regain our footing, news organisations need to be clear about what has gone wrong, what we are doing right, and what those 'rules of the game' should have been in the first place.

Some in the media are now arguing that the problem lies with the very idea that we should aim for objectivity, that this has become a system that allows a select group to determine what 'objective truth' means, or create a formula that blindly defines 'neutrality' as the midpoint in a discussion, even when one side is resorting to lies and misinformation. Those are valid criticisms – of the lack of diversity in newsrooms, and of what some journalists do – but in my view, they completely misunderstand what journalistic objectivity was supposed to mean.

Journalism is not stenography. It was never supposed to involve unthinkingly transcribing whatever nonsense a political leader might spout about the price of meat or takeaway pizzas without making an assessment of those claims. It was never about a reflexive positioning midway between the views of opposing parties. If someone is demonstrably lying, or talking through their hat, then that is what we should say. As Alan Sunderland, journalist and former ABC executive, wrote in *Meanjin* recently in response to such arguments, 'regurgitating the views of others without assessing their factual basis is not journalism. Balancing a smart well-informed view with an ignorant ill-informed view and giving them the same weight is not journalism. Failing to care about where the truth lies is not journalism.'[7]

Sunderland expressed a 'great fear', which I share, that 'deciding that modern journalism is not doing its job properly, the solution will be encouraging it to move away from notions of impartiality and towards being a partisan player in the game, instead of recommitting itself to the clear-headed and powerful role it has always needed to play as an honest broker for the facts.'

An even more powerful exposition of this case came from Tom Rosenstiel – who, along with Bill Kovach, wrote the text *The Elements of Journalism* – in a long Twitter thread[8] responding to a powerful essay in the *New York Times* by the Pulitzer-prize winning reporter Wesley Lowery.[9] I recommend the whole thread, but I'll summarise it here.

Rosenstiel argues that objectivity was never meant to devolve to 'he said/she said' journalism but rather required journalists to employ objective, observable, repeatable methods of verification, precisely because they could never be personally objective.

'I fear a new misunderstanding is taking root in newsrooms today, one that could destroy the already weakened system of journalism on which democracy depends,' he tweeted.

'That misunderstanding is the idea that if we adopt subjectivity to replace a misunderstood concept of objectivity, we will have magically arrived at truth – that anything I am passionate about and believe deeply is a kind of real truth.

'. . . If journalists replace a flawed understanding of objectivity by taking refuge in subjectivity and think their opinions have more moral integrity than genuine inquiry, journalism will be lost.

'If we mistake subjectivity for truth, we will have wounded an already weakened profession at a critical time. If we lose the ability to understand other points of view we will have allowed our passions to overwhelm the purpose democratic society requires of its press.'

Or, to put it the other way round, if we plant our feet firmly on the foundational idea that we follow the facts wherever they lead, we could strengthen our profession at this critical time, when democratic society desperately needs it.

Misinformation and dangerous conspiracy theories thrive when people are uncertain and stressed and alone – exactly the conditions created by this disease, which requires us to isolate from one another to survive. As Wade Davis wrote in August in *Rolling Stone*, 'pandemics and plagues have a way of shifting the course of history.'[10]

When we are finally able to resume our lives, information and considered debate could chart our post-pandemic course for the better. We might manage to reinstate facts to our consideration of climate policy and abandon the notion that caring about the future of the planet is a partisan issue, or something to be exploited in the interests of short-term 'brutal retail politics'.

If we call out lies, are curious and open-minded to different points of view and ideas, and can brace ourselves against those who would turn everything into some kind of 'war', we might help nurse civic debate back to something constructive, as we try to recover and rebuild.

Perhaps, just perhaps, the way we handled the early stages of the pandemic has proved that this is possible.

Lenore Taylor is the *Guardian Australia*'s editor. She has won two Walkley awards and has twice won the Paul Lyneham Award for excellence in press gallery journalism. She co-authored a book, *Shitstorm*, on the Rudd government's response to the global economic crisis.

The plague and the cultivation of an inner life

NYADOL NYUON

In early March, I flew to New Zealand through the busy Tulla-marine Airport. I returned to a country in lockdown. I had been to speak at the New Zealand Festival of the Arts held in Wellington. Life was normal. We moved freely: going out for drinks, eating at various restaurants, hugging friends and shaking hands. We even went to a club to dance. It was packed as sweaty, dancing bodies pumped into each other. We casually spoke about the spread of the coronavirus as it began to emerge as a potentially serious public health issue but the consequences and impact of the disease felt distant. It was still happening far away. It was not yet an issue to worry about or to change one's plans to accommodate. At that time, such a reaction would have appeared exaggerated. The events that followed over the next few days were unimaginable.

At the festival, I had presented to a full room of a few hundred people; twenty-four hours later, that felt like a bygone era. By the time I landed in Melbourne, restrictions were in place and large gatherings had been banned. I went home and began my fourteen days of isolation. It was difficult to keep up with the pace of change. In Victoria, events progressed to a state of emergency. Back in New Zealand, the country went into a nationwide lockdown. The world became a different place within weeks.

The large issues – the politics and economics of the pandemic – were constantly discussed on television and in media releases by state and federal governments. These announcements did little to calm people. There was panic buying that led to supermarkets rationing essential products. The rationing of products reminded me of when I lived in Kakuma Refugee Camp in Kenya before being resettled in Australia. I used to wait for the United Nations food rations, provided fortnightly. The lines for the rations were long, just like the lines of people now waiting to register for welfare payments in suburbs all around Australia.

I saw my old world of Kakuma reflected in the new world enforced by the plague. The old world involved facing up to your life, daily, knowing you had little control. You had to live with uncertainty. The challenge of living with uncertainty, however, is that you never get used to it. The need, the human desire, for returning to something more grounded is always lingering and unsettling. These feelings leave one caught in a state of waiting to begin living again when normalcy returns. We are living in such a world now.

Since the beginning of the pandemic we have been waiting – waiting to see whether we are next, waiting to flatten the curve, waiting for a vaccine, waiting to return to a life before the plague. The wait is not over yet. I write as Victoria returns to the beginning of the wait. We are going back into lockdown for the second time. The situation is no longer an emergency, it is a state of disaster, meaning that the premier of Victoria is satisfied that the state is suffering an emergency that constitutes a significant and widespread danger to life or property.

For many, we cannot wait for this to be over and get back to our lives – maybe even pick things up from where the plague forced a pause and yanked us out of the lives we had built. I have wondered, however, what exactly do I want to go back to? My life before the plague was not perfect, and even if perfection should not be the goal, it was not a lived life, it was a rushed life. A life with a laundry list of things that needed to be ticked off.

I got into the habit of rushing from one thing to another as a way of fulfilling promises I had made to myself when I was a refugee in Kakuma. Now I'm a grown woman, this habit has become a way of life, which was draining my days of joy and putting me on a path that included ticking 'collapsing in a heap' off my list of achievements. The habit started as a plea when I was a desperate teenager with big dreams but was limited by the confines of a refugee camp. Every night, after my mother had finished singing gospel songs and praying, I would lie on my back, stare up into the dark ceiling of the room that my whole family shared, and begin pleading. I vowed to God that in return for getting my family out of the camp, I would be a good Christian and I would take all opportunities that came my way in my new country.

My family arrived in Australia in March 2005. Within a few months I was enrolled in Year Eleven. I grabbed every opportunity and attempted to get the best out of it. I completed two degrees, including a law degree from the University of Melbourne. I achieved a childhood dream when I was admitted to the legal profession as an Australian lawyer in 2016. I got a job on Collins Street, working at Arnold Bloch Leibler, a leading Australian commercial law firm. I became a mother. I took on the new challenge of publicly advocating against racism. I never paused – there was always something to be done and to be done well. The list of things that needed to be done kept growing, and the pace at which they needed to be done kept getting faster – until the plague forced a halt. Many of my commitments were cancelled. It seemed like life itself was being cancelled, as seeing family and friends or attending work was no longer possible. These were the best outcomes: some people lost their lives.

As the days turned to weeks, and weeks to months, in various stages of lockdown, my life became very different from the life I had been living a few months before. The pace of life had slowed down considerably. I began to wonder what was salvageable – even a blessing in disguise – during this unusual time.

The plague is not a beautiful thing, but something can be retrieved even from the worst of circumstances. This is not to minimise experiences of pain, loss, or anguish. There is nothing enlightened about suffering. Surviving a traumatic event is not a prerequisite to a good or reflective life. I know positive thinking does not feed your hunger, soothe pain, or pay bills but neither does becoming depressed over life's difficulties. This is not tough love; it is life that is tough. Many of us know that to be true, but the pandemic is forcing us to reckon with this in a manner that would never have been possible before. How we respond to this new reality will be determined by the individual circumstances of our lives and the privileges we enjoy. I have found those answers that understate or exaggerate the situation to be fruitless.

We have seen the impact of exaggerated reactions because they tend to be public, ugly, and harmful. This includes panic buying that leaves vulnerable members of society without access to essential goods. We have also witnessed the irrational racism that seeks to blame others for causing or spreading the virus. It began with racist abuse and attacks on Asian people living in Australia. It has now, assisted by media reports, spread to other immigrants' communities. A new term, 'COVID racism', has been introduced to describe the old fear of the other.

If the exaggerated reactions are outward, the understated ones tend to be interpersonal and private. We are undoubtedly privileged in Australia, but we cannot, however, understate *how* privileged we are as a global plague takes hold. I have witnessed people whose eyes express concern comforting themselves, and probably others, with the phrase 'But I am so lucky'. I hear this often from friends, and suspect part of the reason they feel the need to say so is because when speaking to someone who they think has survived worse, they want to show compassion. That is good, but it is not necessary. I think it is permissible to feel your emotions as they are without worrying whether the reaction is appropriate when compared to others' past experiences. This is not to reject the value of reflecting. It is to say that our reflection on privilege can turn into self-blame. Why am I not coping well? You are not coping because you are in a fucking global pandemic. Salvage what you can.

What I want to salvage from the wreck left by this pandemic is a fresh point of view and a new way of life. I am borrowing the idea that this unprecedented event is a 'sacred pause'. I do not want to return to the rush of my life as it was before the plague. I want to live. The idea that this is a 'sacred pause', an opportunity to rebuild a life (for those of us who will survive), gives us an opportunity to re-examine our lives while the noise of the world has turned down. Perhaps now we can hear whatever it is that our inner voice has been struggling to tell us as it tried to compete with the buzz of a busy life in a busy world. We have an opportunity to ask whether all the things we used to do, and which we can't do now, brought meaning to our lives. We can now weigh up what truly belongs and what can be left in the life before the plague.

Our resolve to hold on to the reflective view of a 'sacred pause' will be tested as we try to conjure up lives while living in lockdown. There are times I struggle to remain positive. Yes, I have a stable job. Thankfully, my children are well. But I have also missed my brother's surgery, which he went through alone though he lives in Melbourne, no more than thirty minutes' drive from where my children and I moved. So far, I have only spoken to him by phone. I miss him. My children miss him, as I am reminded by my daughter asking when she will see him so they can go to see the dinosaurs. We haven't had the chance to interact with our new neighbours. They have children around my children's age and my daughter is eager to make friends. Yet they have to interact mostly by waving at each other through the window. My daughter often protests, asking why she can't meet the neighbours; I tell her it is because of coronavirus.

Recently I caught her discouraging her brother from playing with one of her toys by telling him, 'It is coronavirus.' She is learning to live with the plague, maybe even to use it to her advantage. That is the task we all face now.

What has been pitilessly clarified by the plague is that there is no limit to what we can lose. Seneca, the Stoic philosopher, wrote: 'Remember that all we have is "on loan" from fortune, which can reclaim it without our permission – indeed, without even advance notice.' When fortune comes for the outer life, we can retreat to the inner life and tap into resources of resilience, courage, and fortitude to keep ourselves 'together' – as Nina Simone said – until things change.

For now, our inner strength does not need to be tested by jumping from planes, climbing Mount Everest or running a marathon. Instead, we have to attempt to survive a global pandemic, mostly at home. To retreat into our inner self, we must cultivate that self.

Perhaps I am putting too much of a positive spin on it, but this is what I hope to learn from the plague. To begin the imperfect process of practicing a more lived life.

Nyadol Nyuon is an Australian lawyer and human rights advocate, who was born in a refugee camp in Ethiopia, of a family fleeing the Second Sudanese Civil War. She works at the law firm Arnold Bloch Leibler in Melbourne and is a regular media commentator. Her book, a collection of essays on race, identity and belonging is coming out with UQP in 2021.

Notes

This place of sickness

1 Timothy Neale, 'What are whitefellas talking about when we talk about "cultural burning"?', Inside Story, 17 April 2020, insidestory.org.au/what-are-whitefellas-talking-about-when-we-talk-about-cultural-burning/, accessed 24 August 2020.

2 Victor Steffensen, *Fire Country: How Indigenous fire management could help save Australia,* Hardie Grant, Melbourne, 2020, p. 91.

3 James Boyce, *1835: The founding of Melbourne and the conquest of Australia,* Black Inc., Melbourne, 2011, p. 197.

Flames

1 Maddie Stone, 'Australia's fires blew smoke 19 miles into the sky, similar to a predicted nuclear blast', *Washington Post*, 23 June 2020, washingtonpost.com/weather/2020/06/22/australias-fires-blew-smoke-19-miles-into-sky-similar-predicted-nuclear-blast/, accessed 15 September 2020.

Drawing breath

1 Inger Andersen and Johan Rockström, 'COVID-19 is a symptom of a bigger problem: our planet's ailing health', *Time*, 5 June 2020, time.com/5848681/covid-19-world-environment-day/, accessed 25 August 2020.

2 David Quammen, 'We made the coronavirus epidemic', *New York Times*, 28 January 2020, nytimes.com/2020/01/28/opinion/coronavirus-china.html, accessed 25 August 2020.

Dead water

1 Sarah Murgatroyd, *The Dig Tree*, Text Publishing, Melbourne, 2002.

2 Robert Glasser, 'Preparing for the Era of Disasters', Australian Strategic Policy Institute, 6 March 2019, aspi.org.au/report/preparing-era-disasters, accessed 25 August 2020.

3 Declan Gooch and Saskia Mabin, 'Darling River flows at Menindee Lakes, washing away spectre of mass fish kills', ABC Broken Hill, 12 March 2020, abc.net.au/news/2020-03-12/water-arrives-at-menindee-lakes-darling-river-fish-kill-nsw-town/12048940, accessed 7 August 2020.

4 'Final Report: Independent assessment of the 2018–19 fish deaths in the lower Darling', Murray-Darling Basin Authority, 29 March 2019, mdba.gov.au/sites/default/files/pubs/Final-Report-Independent-Panel-fish-deaths-lower%20Darling_4.pdf, accessed 7 August 2020.

5 Gooch and Mabin, 'Darling River flows at Menindee Lakes', ABC Broken Hill, 12 March 2020.

6 Paige Cockburn, 'Menindee fish kill leaves devastated town wondering if its future is gone too', ABC News, 20 January 2019, abc.net.au/news/2019-01-20/menindee-fish-death-leaves-devastated-town-worried-about-future/10729132, accessed 7 August 2020.

7 'Water management facts', Murray–Darling Basin Authority, updated 15 May 2020, mdba.gov.au/managing-water/water-management-facts, accessed 7 August 2020.

8 'Assessment of river flows in the Murray–Darling Basin: Observed versus expected flows under the Basin Plan 2012-2019', Wentworth Group of Concerned Scientists, August 2020, wentworthgroup.org/2020/09/mdb-flows-2020/2020/, accessed 15 September 2020.

9 Margaret Simons, 'Govt tight-lipped on ACCC Murray–Darling Basin water report', *Saturday Paper*, 4 July 2020, thesaturdaypaper. com.au/news/politics/2020/07/04/govt-tight-lipped-accc-murray-darling-basin-water-report/159378480010062, accessed 7 August 2020.

10 Jamie Pittock. 'Are we there yet? The Murray–Darling Basin and sustainable water management', Thesis Eleven, 2019, 150:1, p. 120.

11 Bureau of Meteorology, 'State of the Climate 2018' report, bom. gov.au/state-of-the-climate/, accessed 7 August 2020.

12 J. Pittock and M. Colloff, 'Why we disagree about the Murray–Darling Basin Plan: Water reform, environmental knowledge and the science-policy decision context', *Australasian Journal of Water Resources*, 2019, 23:2, pp. 88–98.

13 Julie Power, 'Macquarie marshes: A great sight of nature only being kept alive by life support', *Sydney Morning Herald*, 19 November 2016, smh.com.au/environment/conservation/macquarie-marshes-a-great-sight-of-nature-only-being-kept-alive-by-life-support-20161119-gssyv3.html, accessed 7 August 2020.

14 Aneeta Bhole 'Indigenous community say they've lost their culture to water mismanagement', SBS News, 18 October 2019, sbs.com.au/news/indigenous-community-say-they-ve-lost-their-culture-to-water-mismanagement, accessed 7 August 2020.

A matter of urgency

1 Will Steffen quoted in '"Collapse of civilisation is the most likely outcome": top climate scientists', *Voice of Action*, 5 June 2020, voiceofaction.org/collapse-of-civilisation-is-the-most-likely-outcome-top-climate-scientists/, accessed 25 August 2020.

Don't blink

1 Kai Heron and Jodi Dean, 'Revolution or Ruin', *e-flux*, #110, June 2020, e-flux.com/journal/110/335242/revolution-or-ruin/, accessed 2 July 2020.

Our daily bread

1 Department of Agriculture, Water and the Environment, Australian Bureau of Agricultural and Resource Economics and Sciences (ABARES), 'Snapshot of Australian Agriculture', 2020, agriculture.gov.au/abares/publications/insights/snapshot-of-australian-agriculture#economic-performance-is-driven-by-the-most-productive-farms, accessed 1 July 2020.
2 ABARES, 'Vegetable farming', agriculture.gov.au/abares/research-topics/surveys/vegetables#detailed-physical-characteristics, accessed 1 July 2020.

The music of the virus

1 James Knowlson, *Damned to Fame: The life of Samuel Beckett*, Bloomsbury, London, 1993, p. 66.
2 Knowlson, *Damned to Fame*, p. 95.
3 Samuel Beckett quoted in Knowlson, *Damned to Fame*, p. 546.
4 Sophy Roberts, *The Lost Pianos of Siberia*, Doubleday, London, 2020, p. 28.

One voice

1 Kim Scott, Russell Nelly, Helen Hall and Wirlomin Noongar Language and Stories, *Dwoort Baal Kaat*, University of Western Australia Publishing, Crawley, 2013.
2 Don Watson, 'Leaders and dung beetles', *The Monthly*, February 2020, themonthly.com.au/issue/2020/february/1580475600/don-watson/leaders-and-dung-beetles, accessed 13 July 2020.
3 Michael Dulaney, 'The next pandemic is coming – and sooner than we think, thanks to changes to the environment', ABC

News, 7 June 2020, abc.net.au/news/science/2020-06-07/a-matter-of-when-not-if-the-next-pandemic-is-around-the-corner/12313372, accessed 6 July 2020.

4 Don Watson, 'What lessons will we learn from the virus?', *The Monthly*, May 2020, themonthly.com.au/issue/2020/may/1588255200/don-watson/what-lessons-will-we-learn-virus, accessed 13 July 2020.

5 Thomas Keneally, 'These fires have changed us', *Guardian*, 1 February 2020, theguardian.com/australia-news/2020/feb/01/thomas-keneally-these-fires-have-changed-us, accessed 6 July 2020.

6 Bill Bowtell, 'The risks as lockdowns loosen', *Saturday Paper*, 16–22 May 2020, thesaturdaypaper.com.au/opinion/topic/2020/05/16/the-risks-lockdowns-loosen/15895512009848, accessed 6 July 2020.

7 Gregg Borschmann, Oliver Gordon and Scott Mitchell, 'Rio Tinto blasting of 46,000-year-old Aboriginal sites compared to Islamic State's destruction in Palmyra', ABC News, 29 May 2020, abc.net.au/news/2020-05-29/ken-wyatt-says-traditional-owners-tried-to-stop-rio-tinto-blast/12299944, accessed 6 July 2020.

8 Bill Gammage, *The Biggest Estate on Earth: How Aborigines Made Australia*, Allen & Unwin, Sydney, 2012, pp. 136, 323, 138.

9 Noel Pearson, *A Rightful Place: Race, recognition and a more complete Commonwealth*, Black Inc., Melbourne, 2014, p. 55.

10 'Kukenarup Memorial', Wikipedia, last edited 7 March 2020, en.wikipedia.org/wiki/Kukenarup_Memorial, accessed 6 July 2020.

About the birds this spring

1 David Bowman et al, 'Wildfires: Australia needs national monitoring agency', *Nature*, 12 August 2020, nature.com/articles/d41586-020-02306-4, accessed 25 August 2020.

NOTES

2 Miki Perkins, 'Rare visitor excites Melbourne, yet its presence is
a sad sign', *Sydney Morning Herald*, 30 May 2020, smh.com.au/
environment/climate-change/rare-visitor-excites-melburnians-
yet-its-presence-is-a-sad-sign-20200530-p54xxp.html, accessed
25 August 2020.

3 Craig Butt and Justin McManus, '"It's a sorry sight": Dead birds
wash up on Mallacoota's beaches', *The Age*, 7 January 2020,
theage.com.au/national/victoria/it-s-a-sorry-sight-dead-birds-
wash-up-on-mallacoota-s-beaches-20200107-p53plm.html,
accessed 25 August 2020.

4 'Australian magpie mimics emergency siren during NSW bushfires:
video', *Guardian*, 2 January 2020, theguardian.com/environment/
video/2020/jan/02/australian-magpie-mimics-emergency-siren-
during-nsw-bushfires-video, accessed 25 August 2020.

5 Emily Clark, 'Waking up', ABC News, 1 February 2020, abc.
net.au/news/2020-02-01/natural-bushfire-recovery-underway-
binna-burra,-blue-mountains/11916742?nw=0, accessed
25 August 2020.

6 'Chris Packham's spring: video', BBC *Springwatch* via Facebook,
28 May 2020, facebook.com/watch/?v=295459508286712,
accessed 25 August 2020.

7 Colm Keena, 'Coronavirus: Birdsong seems louder and the
ravens are more relaxed', *Irish Times*, 17 April 2020, irishtimes.
com/news/ireland/irish-news/coronavirus-birdsong-seems-
louder-and-the-ravens-are-more-relaxed-1.4231725, accessed
25 August 2020.

8 'First cuckoo 2020', Bird Spot, birdspot.co.uk/first-cuckoo,
accessed 25 August 2020.

9 Sam Knight, 'Swifts and the fantasy of escape', *New Yorker*, 26 July
2020, newyorker.com/news/letter-from-the-uk/swifts-and-the-
fantasy-of-escape, accessed 25 August 2020.

10 Chris Green, '"You can hear into the distance": Wildlife sound
expert on how coronavirus has changed the world', *iNews UK*,
10 April 2020, inews.co.uk/news/coronavirus-lockdown-

230

wildlife-expert-bird-songs-environment-nature-417130, accessed 25 August 2020.

11 Shannon Mattern, 'Urban auscultation; or, perceiving the action of the heart', *Places Journal*, April 2020, placesjournal.org/article/urban-auscultation-or-perceiving-the-action-of-the-heart/?cn-reloaded=1&cn-reloaded=1, accessed 25 August 2020.

12 Leanne Italie, 'Hopeful bird, foreboding sirens: a pandemic in sounds', NBC Washington, 6 April 2020, nbcwashington.com/news/coronavirus/hopeful-birdsong-foreboding-sirens-a-pandemic-in-sound/2265854/, accessed 25 August 2020.

13 Mark M. Smith, 'Welcome to your sensory revolution, thanks to the pandemic', *The Conversation*, 27 April 2020, theconversation.com/welcome-to-your-sensory-revolution-thanks-to-the-pandemic-136321, accessed 25 August 2020.

14 Michael Boniface, '"Plague doctor" spotted again in Crouch End "standing by the melons" outside Park Road newsagent', *Ham & High*, hamhigh.co.uk/news/crouch-end-plague-doctor-spotted-again-outside-newsagent-1-6643064, accessed 25 August 2020.

15 'Coronavirus: Hellesdon plague doctor given advice by police', BBC News, 4 May 2020, bbc.com/news/uk-england-norfolk-52533718, accessed 25 August 2020.

16 Erin Blakemore, 'Why plague doctors wore those strange beaked masks', *National Geographic*, 13 March 2020, nationalgeographic.co.uk/history-and-civilisation/2020/03/why-plague-doctors-wore-those-strange-beaked-masks, accessed 25 August 2020.

A reinstatement of the facts

1 Lenore Taylor, 'Abbott's beef on carbon price doesn't add up', *Sydney Morning Herald*, 9 April 2011, smh.com.au/politics/federal/abbotts-beef-on-carbon-price-doesnt-add-up-20110408-1d7fw.html, accessed 25 August 2020.

2 Simon Benson and Steve Lewis, 'Businesses forced to dump carbon tax hike on customers', News.com.au, 16 July 2012,

www.news.com.au/finance/small-business/dumped-with-big-carbon-tax-hike/news-story/a3cc18f16e9348dd55f886481753486f?sv=2cc7270d7f4daabea47a15fae35d28b, accessed 25 August 2020.

3 Lenore Taylor, 'Labor boxed in over the price of pies and pizzas', *Sydney Morning Herald*, 21 July 2012, smh.com.au/politics/federal/labor-boxed-in-over-the-price-of-pies-and-pizzas-20120720-22f6p.html, accessed 25 August 2020.

4 'Carbon tax just brutal politics: Credlin', SBS News, 12 February 2017, sbs.com.au/news/carbon-tax-just-brutal-politics-credlin, accessed 25 August 2020.

5 Malcolm Farr, 'Australians' trust in government and media soars as coronavirus crisis escalates', *Guardian*, 7 April 2020, theguardian.com/australia-news/2020/apr/07/australians-trust-in-government-and-media-soars-as-coronavirus-crisis-escalates, accessed 25 August 2020.

6 Simon Jackman and Shaun Ratcliff, 'America's Trust Deficit', United States Studies Centre, University of Sydney, 18 February 2018, ussc.edu.au/analysis/americas-trust-deficit, accessed 25 August 2020.

7 Alan Sunderland, 'A partiality for the truth', *Meanjin*, 12 December 2019, meanjin.com.au/blog/a-partiality-for-the-truth/, accessed 25 August 2020.

8 Tom Rosenstiel, 'I'm not avid on Twitter but ...', Twitter, 24 June 2020, twitter.com/tomrosenstiel/status/1275773988053102592, accessed 25 August 2020.

9 Wesley Lowery, 'A reckoning over objectivity, led by black journalists', *New York Times*, 23 June 2020, nytimes.com/2020/06/23/opinion/objectivity-black-journalists-coronavirus.html, accessed 25 August 2020.

10 Wade Davis, 'The Unraveling of America', *Rolling Stone*, 6 August 2020, rollingstone.com/politics/political-commentary/covid-19-end-of-american-era-wade-davis-1038206/, accessed 28 August 2020.

Acknowledgements

This anthology doesn't just reflect the work of twenty-five fine writers, but of three organisations – the Copyright Agency, the *Guardian* and Penguin Random House Australia, all of whom were keen to both document this extraordinary year, and to support Australian literature through that year. My gratitude to Nicola Evans and the Copyright Agency, who had the idea of producing this anthology; to publisher Meredith Curnow, who embraced and supported that idea, and to the *Guardian*'s Lucy Clarke who published the essays along the way. Thank you all for helping select the writers, a difficult job given that we were spoilt for choice, and, of course, for asking me to be a part of your team. – S.C.